Myxobacteria: Physiology and Regulation

Myxobacteria: Physiology and Regulation

Editor

David Whitworth

MDPI • Basel • Beijing • Wuhan • Barcelona • Belgrade • Manchester • Tokyo • Cluj • Tianjin

Editor
David Whitworth
Institute of Biological,
Environmental and Rural
Sciences Aberystwyth University
Aberystwyth
UK

Editorial Office
MDPI
St. Alban-Anlage 66
4052 Basel, Switzerland

This is a reprint of articles from the Special Issue published online in the open access journal *Microorganisms* (ISSN 2076-2607) (available at: https://www.mdpi.com/journal/microorganisms/special_issues/Myxobacteria).

For citation purposes, cite each article independently as indicated on the article page online and as indicated below:

LastName, A.A.; LastName, B.B.; LastName, C.C. Article Title. *Journal Name* **Year**, *Volume Number*, Page Range.

ISBN 978-3-0365-3924-9 (Hbk)
ISBN 978-3-0365-3923-2 (PDF)

Contents

About the Editor

David Whitworth is Professor of Biochemistry at Aberystwyth University. He obtained a B.A. in Biochemistry from Oxford University and went on to study for a Ph.D. in molecular microbiology at Warwick University under the supervision of Prof David Hodgson. His Ph.D. thesis was on light-dependent carotenoid production in *Myxococcus xanthus* and he has continued to work on prokaryotic signalling pathways and myxobacteria since then. After periods as a lecturer and as a senior research fellow at Warwick University, he moved to Aberystwyth University in 2008.

microorganisms

Editorial

Myxobacteria: Physiology and Regulation

David E. Whitworth

Institute of Biological, Environmental and Rural Sciences, Aberystwyth University, Aberystwyth SY23 3DD, UK; dew@aber.ac.uk

Myxobacteria are fascinating and important prokaryotes. Their genomes are exceptionally large, endowing them with a wide range of interesting behaviors, including (but certainly not limited to) multicellular fruiting body formation, gliding motility, social interactions, predation, and secondary metabolite production. Their ecological importance in major ecosystems is becoming established, and we are beginning to understand the evolutionary forces that have shaped their current phenotypes and behaviors. Novel species of myxobacteria are steadily being discovered, often producing unusual metabolites and enzymes. There is also significant biotechnological interest in these organisms for a wide range of potential applications. Molecular studies, ranging in subject from individual enzymes to entire 'omes, continue to provide rich insights into myxobacterial biology.

This Special Issue brings together five research articles and three reviews, to provide a snapshot of current myxobacterial research in all its diversity. We are very grateful to the authors for submitting their work to the issue.

Myxobacteria are well-known for their multicellular behaviors which are often associated with morphological changes [1]. When starved of nutrients, coordinated motility of cells in the population results in the formation of ripple patterns and multicellular aggregates, which develop into fruiting bodies. Inside the fruiting bodies, rod-shaped cells sporulate and remain dormant as round myxospores until nutrients become available once more. Zhang, H. et al., have provided a review of morphological changes during sporulation and germination, focusing on changes in peptidoglycan synthesis and degradation during the sphere-to-rod transition. In contrast, Zhang, J. et al., investigated population-level morphology, providing a novel computational approach for combining information from multiple types of microscopy.

Citation: Whitworth, D.E. Myxobacteria: Physiology and Regulation. *Microorganisms* **2022**, *10*, 805. https://doi.org/10.3390/microorganisms10040805

Received: 25 March 2022
Accepted: 12 April 2022
Published: 12 April 2022

The gene regulatory network underpinning fruiting body formation in myxobacteria is highly complex and it has taken decades of molecular genetics experiments to achieve our current understanding of the network [2]. Whitworth and Zwarycz surveyed regulatory genes in myxobacterial genomes to assess the conservation of signalling genes, also providing insights into how evolution has molded contemporary myxobacterial regulatory networks. Their analysis extended to quorum signaling systems and the Car system, which regulates the light-dependent production of photoprotective carotenoids. In their review, Padmanabhan et al., provide a thorough and mechanistically detailed description of the Car system-a particularly well-understood gene-regulatory network system which has also provided paradigms for widely distributed protein families with diverse functions.

Myxobacterial predation has received increasing research attention in recent years [3]. In this Special Issue, two research articles address aspects of the evolution and ecology of predation. Nair and Velicer used co-evolution experiments to investigate the evolutionary relationships between myxobacteria and their prey bacteria. They found that myxobacterial predators promote long-term diversity within prey communities. Mayrhofer et al., investigated the trophic interactions between nematodes, myxobacteria and prey bacteria, observing that myxobacteria and nematodes affect each other's predatory behavior towards their shared prey.

Myxobacterial predation involves the secretion of secondary metabolites encoded by biosynthetic gene clusters, many of which are known to have antibiotic activity. Ahearne et al., undertook a genomic analysis of selected myxobacterial strains to assess their taxonomy and determine their biosynthetic potential, demonstrating that that genome-led classification can help prioritize natural product discovery. In their review of myxobacterial genomics, Whitworth et al., also focused on genome analysis as an aid for taxonomic classification, and showed how genomic information has enabled the application of post-genomic technologies to diverse aspects of myxobacterial research, including secondary metabolite biosynthesis.

Funding: This research received no external funding.

Conflicts of Interest: The author declares no conflict of interest.

References

1. Muñoz-Dorado, J.; Marcos-Torres, F.J.; García-Bravo, E.; Moraleda-Muñoz, A.; Pérez, J. Myxobacteria: Moving, Killing, Feeding, and Surviving Together. *Front. Microbiol.* **2016**, *7*, 781. [CrossRef] [PubMed]
2. Kroos, L. Highly Signal-Responsive Gene Regulatory Network Governing Myxococcus Development. *Trends Genet.* **2017**, *33*, 3–15. [CrossRef] [PubMed]
3. Whitworth, D.E.; Jurkevitch, E.; Pérez, J.; Fuhrmann, G.; Koval, S.F. Editorial: Mechanisms of Prokaryotic Predation. *Front. Microbiol.* **2020**, *11*, 2071. [CrossRef] [PubMed]

microorganisms

MDPI

Review

Myxococcus xanthus as a Model Organism for Peptidoglycan Assembly and Bacterial Morphogenesis

Huan Zhang, Srutha Venkatesan and Beiyan Nan *

Department of Biology, Texas A&M University, College Station, TX 77843, USA; hzhang@bio.tamu.edu (H.Z.); svenkatesan@bio.tamu.edu (S.V.)
* Correspondence: bnan@tamu.edu; Tel.: +1-979-845-3487

Abstract: A fundamental question in biology is how cell shapes are genetically encoded and enzymatically generated. Prevalent shapes among walled bacteria include spheres and rods. These shapes are chiefly determined by the peptidoglycan (PG) cell wall. Bacterial division results in two daughter cells, whose shapes are predetermined by the mother. This makes it difficult to explore the origin of cell shapes in healthy bacteria. In this review, we argue that the Gram-negative bacterium *Myxococcus xanthus* is an ideal model for understanding PG assembly and bacterial morphogenesis, because it forms rods and spheres at different life stages. Rod-shaped vegetative cells of *M. xanthus* can thoroughly degrade their PG and form spherical spores. As these spores germinate, cells rebuild their PG and reestablish rod shape without preexisting templates. Such a unique sphere-to-rod transition provides a rare opportunity to visualize de novo PG assembly and rod-like morphogenesis in a well-established model organism.

Keywords: germination; bacterial cell wall; sporulation; germination; morphology

Citation: Zhang, H.; Venkatesan, S.; Nan, B. *Myxococcus xanthus* as a Model Organism for Peptidoglycan Assembly and Bacterial Morphogenesis. *Microorganisms* **2021**, 9, 916. https://doi.org/10.3390/microorganisms9050916

Academic Editor: David Whitworth

Received: 22 March 2021
Accepted: 22 April 2021
Published: 24 April 2021

1. Introduction

While spheres may seem simple and physically preferable, cells are seldom spherical. Instead, most cells invest extra energy to establish and maintain non-spherical shapes through a process known as "morphogenesis". How cells use molecules at the nanometer scale to establish a defined morphology at micrometer scale has become a fundamental question in biology. Because of their relative simplicity, bacteria are excellent models for studying how genes and proteins determine cell morphology. Rods are the simplest non-spherical shapes adopted by many bacteria. Phylogenic studies suggest that the common ancestor of bacteria was rod-shaped, and that rod-like shapes are advantageous for cell survival [1–3]. Thus, the switch from primeval spheres to rods may mark the origination of bacteria, and represents a landmark morphological transition in evolution. Understanding this switch will likely uncover fundamental mechanisms of morphogenesis.

The word "morphogenesis" (from the Greek words "morphê" and "genesis") means "the beginning of shape". However, despite seeing "the surprising and bewildering riot of shapes" in bacteria [3], true morphogenesis is seldom studied, because we can rarely see the beginning of shape. In most bacteria, rigid peptidoglycan (PG) structures largely determine cell shape. Disruption of PG usually results in the loss of defined cell shape and, eventually, cell death [4,5]. Thus, in order to maintain their shape during growth, bacterial cells must add new PG subunits into existing PG structures [6–9]. When cells divide, the shapes of the daughter cells are already predetermined by their mothers. The preexistence of PG has become a central challenge that impedes progress in understanding bacterial morphogenesis.

One way to circumvent this dilemma is to investigate how cells establish non-spherical shapes from PG-deficient spheres. Some bacteria can shed PG, thus losing their original morphology to cope with environmental stresses, especially when attacked by host immune systems and antimicrobial agents [10,11]. Among these PG-deficient cells, spheroplasts and L-forms derived from rod-shaped cells are able to regenerate rod shapes [11,12]. Importantly, a stable L-form state can be induced in many bacteria by genetically inhibiting the synthesis of PG precursors or upregulating membrane biosynthesis [13–15]. L-form cells of both *Bacillus subtilis* and *Escherichia coli* can rebuild rods when their genetic defects are rescued, which provides potential vehicles for the study of the de novo generation of non-spherical shapes [16,17]. However, these PG-deficient cells usually take on an array of irregular shapes, especially during proliferation, which makes reproducible live-cell imaging technically challenging [16–18]. Moreover, spheroplasts and L-form cells take several generations to restore their original cell shapes, which implies the involvement of cell division in morphogenesis [12,17]. These temporal steps make it difficult to track morphological changes in single cells and to attribute them to a simple PG assembly system. Therefore, in order to better understand bacterial morphogenesis, it would be very helpful to find a system in which homogeneous spheres generate non-spherical shapes independent of both division and pre-existing templates.

Myxococcus xanthus is a Gram-negative bacterium that has been studied extensively for its motility, multicellular development, and predatory behaviors. Vegetative *M. xanthus* cells are long rods (1 µm in diameter and 5–10 µm in length) that go through cell cycles similar to other rod-shaped model organisms, such as *E. coli* and *B. subtilis* [19,20]. In response to certain chemicals, individual *M. xanthus* cells can degrade their PG thoroughly and form spherical spores [21,22]. During germination, these spores restore vegetative morphology by assembling rod-shaped PG de novo. As PG is usually essential for bacterial survival, and its assembly systems are well conserved in most bacteria, including *M. xanthus* [23], *M. xanthus* provides unique opportunities to better understand the mechanisms of PG assembly and rod-like morphogenesis.

2. PG and Cell Shape

The entire PG layer is a continuous, mesh-like macromolecule of glycan strands that are crosslinked via short peptides [24]. PG surrounds the whole cell, provides major mechanical support against turgor pressure, and defines cell shape throughout the life cycles of most bacteria [25]. As PG is chemically unique, and usually essential for cell survival, the synthesis and turnover of PG have been predominant targets for antibacterial treatments [26].

Cells synthesize lipid II, the PG subunit carried by lipids, using conserved enzymes in the cytoplasm and the cell membrane. Lipid II is flipped into the periplasm, where it is assembled into the existing PG structure [9]. PG assembly during vegetative growth relies on two conserved polymerization systems: the Rod complex and class A penicillin-binding proteins (aPBPs). The central components of Rod complex are RodA, PBP2 and MreB. RodA, a SEDS (shape, elongation, division, and sporulation) family transglycosylase (TGase), catalyzes the formation of glycosidic bonds in the glycan strands. PBP2, a member of the class B penicillin-binding proteins (bPBPs) that has transpeptidase (TPase) activity, crosslinks the short peptides on adjacent glycan strands [23,27–30]. MreB, an actin-like cytoskeletal protein that is conserved in most rod-shaped bacteria, is proposed to form the scaffolds that orchestrate PG assembly [31]. Different from the Rod components, aPBPs have both TGase and TPase activity. While the Rod system is essential for rod-like morphology [31–33], the functions of aPBPs are still not fully understood [34]. A recent work suggests that aPBPs contribute to shape maintenance indirectly by repairing PG defects [35]. In general, the Rod complexes reduce cell diameter, whereas aPBPs increase it [36]. Hydrolases also play important roles in PG growth, by generating small openings in the existing PG layer to allow for the insertion of new subunits. Thus, PG is a dynamic structure under the coordinated control of its polymerases and hydrolases [37].

3. The Sporulation and Germination of *M. xanthus*

Many bacteria form spores in order to survive under unfavorable environmental conditions. The endospores formed by Firmicutes such as Bacilli and Clostridia have been subject to extensive studies. In these organisms, the morphological differentiation from rod-shaped vegetative cells to oval spores begins with an asymmetric division, resulting in the formation of a smaller cell—the forespore—and a larger mother cell. Eventually, the forespore is engulfed by, and becomes wholly contained within, the mother cell. The resulting endospores contain intact and, in many cases, thickened PG. In addition, because the cell poles in forespores are generated through division, the mature endospores are also likely to preserve the information of cell polarity [38–40].

Strikingly different from endospore-forming bacteria, cell division is not involved in the sporulation of *M. xanthus*. *M. xanthus* can form spores using two distinct mechanisms, both of which exhibit the transformation of entire rod-shaped vegetative cells into spheres. First, in response to starvation, large groups of vegetative cells can aggregate on solid surfaces and build fruiting bodies, which are filled up with spores. Such spores are difficult to study because in order to release individual spores, the coats of fruiting bodies have to be broken and removed via intensive sonication [20]. In the second mechanism, individual vegetative cells can form dispersed, spherical spores in response to various chemical signals, such as glycerol, dimethyl sulfoxide (DMSO), and agents that inhibit PG synthesis or disrupt PG—including β-lactams, D-cycloserine, fosfomycin, and lysozyme [41,42]. In contrast to fruiting bodies that require millions of cells, solid surfaces, and days to form, chemical-induced sporulation can occur at low cell density, in liquid media, and in a few hours. For example, adding 0.5–1 M glycerol into rich liquid media induces the transition of *M. xanthus* from rod-shaped vegetative cells to spherical spores in 1–3 h [41]. These "quick" spores used to be considered artificial. However, we believe that chemical-induced sporulation is a naturally occurring process through which dispersed cells form spores rapidly under particular environmental stresses. First, many signals that induce quick sporulation are also present in the natural habitats of *M. xanthus*. Second, chemical-induced spores show typical characteristics of starvation-induced spores, such as considerable resistance to heat, desiccation, UV irradiation, sonication, detergents, and enzymatic digestion [43].

Unlike endospores, glycerol-induced *M. xanthus* spores are PG-deficient. Using high-performance liquid chromatography (HPLC) and transmission electron microscopy (TEM), Bui et al. first reported that such spores contained no detectable muropeptides [21]. This conclusion was later confirmed using cryo-electron tomography (cryo-ET) [22]. Thus, during sporulation, vegetative cells thoroughly degrade their PG, shrink into near-perfect spheres [21], and synthesize spore coats that consist of polysaccharides and proteins [44]. Sensing certain environmental signals, such as inorganic ions HPO_4^{2-}, Mg^{2+}, Ca^{2+}, and NH_4^+, *M. xanthus* spores regenerate vegetative cells with rod-like morphology [45]. So far, Ca^{2+} has been found to be a strong germinant that induces roughly synchronized germination [46]. When incubated in rich liquid media containing Ca^{2+}, *M. xanthus* spores can restore vegetative cell shape within 2–3 h [22,46,47].

4. PG Assembly and Morphological Transition during *M. xanthus* Spore Germination

Do *M. xanthus* spores preserve PG subunits from previous vegetative cells? Bui et al. proposed that glycerol-induced *M. xanthus* spores may contain PG subunits [21]. Consistent with this report, these spores are able to germinate and restore vegetative morphology in the presence of fosfomycin, an antibiotic that inhibits the production of UDP-MurNAc—a precursor of PG. However, after the exhaustion of preserved PG subunits, nascent vegetative cells become sensitive to fosfomycin, and are unable to elongate further [22,48]. These results indicate that glycerol-induced spores preserve most of the PG subunits from previous vegetative cells. Thus, PG assembly, rather than the production of PG subunits, is the decisive step for the restoration of rod shape.

Similar to endospores [49], *M. xanthus* spores are refractile, and appear bright under a phase-contrast microscope [46,47]. Upon being suspended in rich media containing Ca^{2+}, most spores lose their refractility within a few minutes, indicating that germination initiates immediately upon induction [46,47]. However, germinating spores remain spherical, and do not elongate until 45–60 min later [22]. Despite the absence of morphological changes, PG assembly initiates immediately once germination starts, as germinating spores begin to incorporate TAMRA 3-amino-D-alanine (TADA)—a fluorescent D-amino acid (FDAA)—evenly onto their surfaces (Figure 1A) [22]. Because FDAAs specifically label newly assembled PG [50,51], the incorporation pattern of TADA indicates that spores first synthesize spherical PG layers in this early phase of germination.

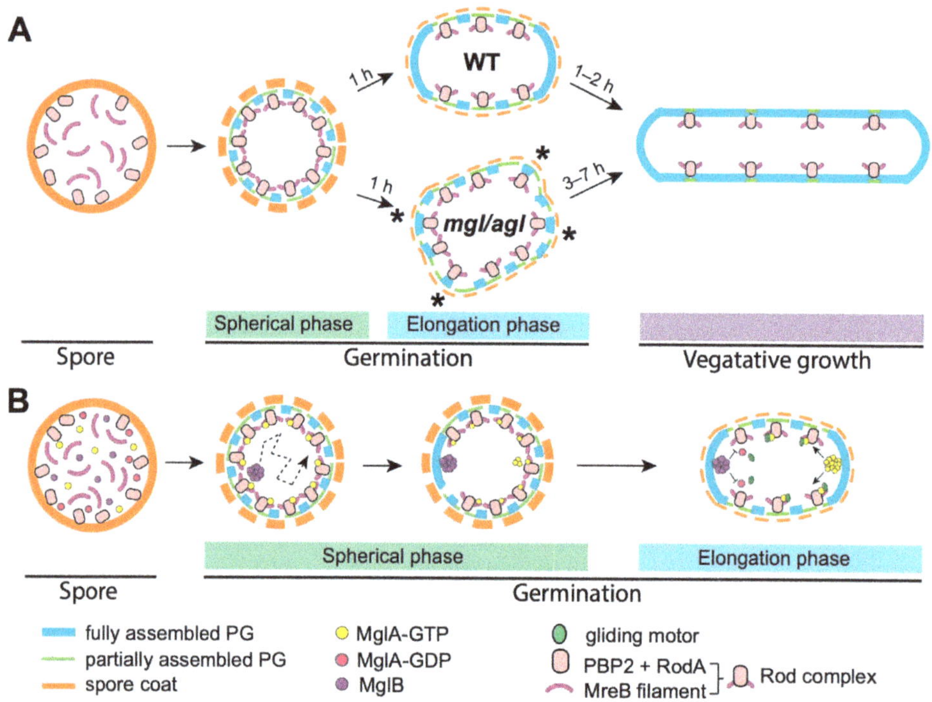

Figure 1. A schematic model for the de novo establishment of rod-shape from spherical, PG-deficient spores. (**A**) As the wild-type (WT) spores germinate, nascent cells restrict Rod complexes, and thus PG assembly, to non-polar regions in the elongation phase of germination. Such a pattern of PG growth maintains the integrity of cell surfaces and restores vegetative morphology within 3 h. In contrast, in the *mgl* spores (from the strains of Δ*mglA*, Δ*mglB*, and $mglA^{Q82L}$ that encode a GTPase-inactive variant of MglA, and $mglB^{OE}$ that overexpresses MglB) and the *agl* spores that express truncated gliding motors, the localizations of Rod complexes and PG assembly are not restricted. As a result, such spores grow into bulged intermediates that are sensitive to osmotic stresses and restore vegetative cell shape much more slowly (4–7 h). The asterisks mark the sites where Rod complexes mislocalize and bulges appear. (**B**) Symmetry breaking by MglB and MglA. The GTP-bound form of MglA (MglA-GTP) connects to Rod complexes via MreB filaments. Through the mutual expulsion between MglB and MglA-GTP, MglB clusters survey the status of PG synthesis indirectly, and cannot localize at the sites where PG assembly is active and Rod complexes are present. Therefore, MglB clusters move randomly in the early spherical phase, when Rod complexes distribute randomly on cell surfaces. Once a patch of PG is completely assembled and the Rod complexes leave, the MglB cluster will stall at this site, which will become the first future pole. At the first pole, MglB expels MglA-GTP, and thus the Rod complexes, toward the second pole. MglA-GTP then occupies the second pole, stimulating the assembly of the gliding machineries, which transport MreB filaments, together with Rod complexes, toward the first pole. As a result, the diametrically opposing clusters of MglA-GTP and MglB restrict the Rod system, and thus the assembly of PG, to non-polar locations.

After remaining spherical for approximately one hour, germinating spores quickly start to elongate. Compared to vegetative cells, which double their length in about four hours, the elongation of germinating spores proceeds quite rapidly, growing one cell length within an hour [22]. During elongation, cells only incorporate TADA to the cylindrical, non-polar portion of their surfaces, indicating that the cell poles become inert for PG growth (Figure 1A) [22].

These distinct growth patterns allow us to divide *M. xanthus* spore germination into two phases: the spherical phase (Phase I), in the first hour of germination, when spores assemble PG evenly on their entire surfaces; and the elongation phase (Phase II), in the second to third hours, when PG growth at the non-polar regions drives cell elongation (Figure 1A). The correlation between the shapes of cells and the patterns of PG growth suggests that a major switch in the mode of PG assembly results in the de novo establishment of rod shape.

5. Roles of aPBPs and the Rod System during Germination

The roles aPBPs and the Rod system play in PG assembly can be studied by monitoring TADA incorporation in the presence of antibiotics, which specifically inhibit either aPBPs or the Rod system. In the spherical phase, neither mecillinam—an inhibitor of PBP2 in the Rod system—nor cefsulodin or cefmetazole—antibiotics that mainly inhibit aPBPs—are able to block TADA incorporation. However, spores stop incorporating TADA when treated with all three antibiotics. In contrast, once spores enter the elongation phase, mecillinam alone is sufficient to block TADA incorporation, whereas cefsulodin and cefmetazole do not show significant effects [22]. *M. xanthus* spores are able to germinate into rods in the presence of cefsulodin and cefmetazole, indicating that aPBPs are dispensable for cell elongation [22]. However, elongated cells revert to spheres after prolonged inhibition of aPBPs, which suggests that aPBPs stabilize rod shapes [22]. This observation echoes the reports on *E. coli*, where aPBPs did not determine rod shape, but rather maintained PG integrity, especially across different pH environments [35,52]. Whereas both aPBPs and the Rod system participate in the assembly of spherical PG layers, the Rod system is essential for the establishment of rod shape.

The Rod proteins are stably produced, and remain active during the entire germination process [22,44], whereas the spatial distribution of the Rod complexes alters dramatically when the spores start to elongate. While RodA localizes randomly on spore surfaces in the spherical phase, it concentrates at non-polar regions during elongation [22]. Similarly to RodA, MreB first appears diffusive in the spherical phase, and then forms filaments that mainly localize at non-polar regions as nascent cells elongate [22]. Such non-polar localization of Rod complexes is consistent with the lateral growth patterns of PG. Taken together, to establish rods from spheres, germinating *M. xanthus* spores must first establish cell poles, and then restrict Rod complexes to non-polar regions.

6. De Novo Establishment of Cell Poles

While *M. xanthus* spores are approximately spherical, those spheres are rarely "perfect". If such morphological "imperfection" preserves polarity from previous vegetative cells, each germinating spore is expected to elongate along the long axis of the ovoid. However, the elongation axis during germination appears random in many spores, independent of their original morphology [22].

Then how do spores establish cell poles de novo? In vegetative *M. xanthus* cells, directed motility requires a head–tail polarity axis, which is under the modulation of multiple regulators, including the Mgl regulators (MglA, MglB and MglC) [53–55], the RomR–RomX pair [56–58], and PlpA [59]. Among these regulators, MglA—a Ras-like GTPase—and its GTPase-activating protein—MglB—are required for rapid cell elongation in the elongation phase of germination. Specifically, fine-tuned MglA is critical for optimal germination efficiency. Deleting MglA or MglB, nullifying the active site of MglA, or overstimulating the GTPase activity of MglA by overproducing MglB all cause severe delays in cell elongation (Figure 1A) [22]. Strikingly, unlike the wild-type spores, which maintain relatively smooth cell surfaces throughout the germination process, these Mgl mutant spores germinate into bulged, multipolar intermediates (Figure 1A). Many of these intermediates lyse before becoming vegetative cells, especially under osmotic stress [22]. Such morphological abnormality directly reflects the structural defects in newly assembled PG. A large portion of Rod complexes heavily incorporate TDA, mislocalize at cell poles, and bulge in elongating Mgl mutant spores. This is in contrast to wild-type spores in which Rod complexes, and therefore PG assembly, are restricted to non-polar regions in the elongation phase of germination (Figure 1A) [22]. These findings indicate that the Mgl system plays a central role in expelling PG assembly from cell poles during germination.

7. Symmetry Breaking by Random Fluctuation

Does polarity originate from preserved spatial cues, or from stochastic fluctuations? During sporulation *M. xanthus* retains both MglA and MglB in its spores [22,60,61]. Here, fluorescence-labeled MglB is seen to form a single bright cluster in each spore throughout the entire course of germination. These clusters first move randomly in the spherical phase, and then abruptly stop moving. Importantly, once an MglB cluster stalls, the spore immediately starts to elongate, using the position of the MglB cluster as one cell pole (Figure 1B) [22]. In contrast, MglA only begins to form a cluster after MglB stalls, and the formation of MglA clusters requires MglB. Once formed, clusters of MglB and MglA always occupy opposite sides of the same spore (Figure 1B) [22]. Thus, the sequential stabilization of MglB and MglA clusters establishes the polarity axis for PG assembly.

The random "walk" of MglB clusters during the spherical phase of germination suggests that polarity is not preserved. The localizations of stalled MglB clusters do not correlate with the geometry of the spores. Instead, MglB clusters are immobilized by the inhibitors of PG assembly—especially the agents that inhibit the Rod system, such as mecillinam [22]. Thus, MglB clusters could stall at the sites where PG assembly has been completed, or has not yet initiated. In this case, it is unlikely that PG assembly has not yet initiated, because no obvious PG assembly activity is observed near the nascent cell poles after the stall of MglB clusters (Figure 1A) [22]. MglB activates the GTPase activity of MglA, and turns MglA-GTP into MglA-GDP. As MglA-GDP cannot form clusters, clusters of MglB and MglA (MglA-GTP) always maintain the farthest distance possible in the same cell [53,54]. As a consequence, once an MglB cluster stalls at one pole, the expulsion between MglB and MglA-GTP causes MglA-GTP to cluster at the opposite side of the spore—the second cell pole (Figure 1B).

8. When PG Assembly Intersects with Gliding Motility

MglB clusters must have the ability to survey the random fluctuation of PG growth during the spherical phase of germination, and stall at the region where PG growth completes first. Once polarity is established, Mgl regulators must expel the Rod complexes from the cell poles. How do Mgl regulators connect to the Rod system? Besides being a component in the Rod system, MreB also supports the gliding motility of *M. xanthus*. MglA-GTP stimulates the assembly of the gliding machineries through direct interaction with MreB [62–65], and then directs them toward non-polar regions [66]. The gliding motors then carry MreB filaments as they move rapidly in the membrane [48,67,68]. MglA-GTP clusters therefore co-localize with MreB filaments that lso carry Rod complexes [53,54,63,69]; however, MglB clusters do not (Figure 1B).

Through the mutual expulsion between MglB and MglA-GTP, MglB clusters stall at the sites where PG assembly is complete and Rod complexes are absent. At the first pole, which contains the MglB cluster, MglB expels MglA-GTP, and thus the Rod complexes, toward the second pole. MglA-GTP then occupies the second pole and stimulates the assembly of the gliding machineries [63,70], which transport the Rod complexes toward the first pole [48,66]. As a result, the diametrically opposing clusters of MglA-GTP and MglB restrict the Rod system, and thus the assembly of PG, to non-polar locations (Figure 1B).

9. Perspectives

9.1. What Does It Take to Make a Rod?

As the assembly of PG is widely conserved in bacteria, the mechanisms by which *M. xanthus* rebuilds rods from spheres might reveal the common principles for rod-like morphogenesis. During *M. xanthus* spore germination, the Rod system is the only element known to be essential for the establishment of rod shape. Similarly, to our knowledge, artificial spheres from other rod-shaped cells are not able to restore their original shapes in the absence of the Rod system [71]. Given their critical roles, we hypothesize that restricting Rod complexes to a non-polar, but expanded region, might be a common prerequisite for building and maintaining a rod. On the one hand, when this restriction is relieved in the Mgl mutants, the cells generate poles randomly and display bulged morphology. On the other hand, in bacteria that over-restrict Rod complexes to very narrow regions, the cells are naturally oval [72,73].

9.2. The Versatile MreB

In comparison to its analogs in other bacteria, MreB in *M. xanthus* is truly extraordinary for its connection to the Mgl regulators and gliding machineries. MreB filaments in *M. xanthus* display a rapid, directed motion that is not yet reported in other organisms [48]. On the other hand, as in most rod-shaped bacteria, *M. xanthus* MreB plays conserved roles in PG assembly. As MreB filaments are intrinsically curved, and bind to the cytoplasmic membranes, the balance between filament bending and membrane deformation can lead MreB filaments to localize at inwardly curved regions [31,74–78]. This localization preference, in turn, affects the localization and dynamics of Rod complexes, which could be sufficient for the maintenance of rod shape [31–33,71,76,79,80]. Additionally, the composition and fluidity of the cytoplasmic membrane could also modulate the localization and dynamics of MreB [81–84]. For instance, rafts of anionic phospholipids preferentially enrich MreB monomers at the cell poles, and expel MreB filaments to non-polar regions [81]. Using these mechanisms, cells are able to generate rod shape spontaneously, based on random fluctuations, albeit much slower. For example, the Mgl and motor mutant spores can still elongate into rods and correct morphological defects after prolonged germination, and the artificial spheres of *E. coli* and *B. subtilis* can regain rod shapes after several generations [17,22]. In addition, Mgl and gliding motors are dispensable for vegetative growth [67,85]. Thus, rather than being the determinants of symmetry breaking, Mgl and gliding motors are supplementary accelerators. Nevertheless, such accelerators provide critical advantages for the survival of *M. xanthus* spores. As chemical-induced *M. xanthus* spores are dispersed and PG deficient, they are vulnerable to biotic and abiotic environments during germination. Equipped with the Mgl regulators and gliding machineries, *M. xanthus* spores are able to regain fitness within one generation.

While the Rod complexes move relatively slowly, with nm/s velocities, gliding motors in *M. xanthus* move significantly faster, at µm/s [48,86–88]. Then how do MreB filaments co-ordinate multiple functions that require distinct dynamics? Aside from PG assembly, MreB affects a broad range of cellular functions, either directly or indirectly—such as membrane organization, DNA replication and segregation, twitching motility, and pathogenesis [89–96]. Studying MreB in *M. xanthus* provides an opportunity to understand how MreB organizes multiple functions simultaneously.

9.3. Are Small GTPases the Universal Regulators of Cell Polarity?

MglA-like GTPases distribute widely in phylogenically diverse bacteria. MglA and its eukaryotic homologs are proposed to have evolved from a common ancestor [97]. GTPase-mediated cell polarization is common in eukaryotes. The Rho-family GTPase Cdc42 and its homologs widely exist, in organisms ranging from yeast to human beings [98]. The rod-shaped yeast *Schizosaccharomyces pombe* forms spherical spores. During germination, Cdc42 first moves randomly during the isotropic growth phase, before stalling at a future pole [98,99]. Analogous to the connection between MglA and the Rod system in *M. xanthus*,

Cdc42 in *S. pombe* is able to survey the integrity of the spore cell wall by interacting with cytoskeletons, cell wall-related enzymes, and molecular motors [99]. Such striking similarities suggest that *M. xanthus* might preserve a prototype of a polarity regulation system that evolved before the divergence between prokaryotes and eukaryotes. Therefore, studying the interactions between Mgl regulators, MreB, and motor-associated proteins may also aid in the understanding of cell polarity in a broad range of organisms.

9.4. What Are the Primary Roles of the Gliding Motors?

Vegetative *M. xanthus* cells move on surfaces using both gliding and twitching motility. In contrast to twitching—which exists in many phylogenically diverse bacteria—gliding, driven by fluid motor complexes, is rather unique to Myxococcales [100,101]. While twitching is required for group behaviors such as coordinated migration, predation, and multicellular development—to name but a few—gliding is rather secondary for vegetative growth [19,102]. Besides accelerating germination, the gliding machineries are also the distributors of spore coat polysaccharides during sporulation, which is critical for the resilience of the spores [22,103]. Thus, rather than facilitating cell locomotion, the primary function of the gliding machineries might be the distribution of various protein complexes in the cell membranes.

9.5. How Do PG Polymerases Co-Ordinate with Hydrolases?

It is commonly accepted that the insertion of new PG subunits is associated with the local hydrolysis of the existing PG network [9,25,37]. It is therefore reasonable to hypothesize that a regulated co-ordination exists between PG polymerases and hydrolases. Hence, exploiting the synergy between these two types of enzymes could usher in new treatments for bacterial infections [104,105]. As of now, our understanding of PG hydrolases has long been hampered by several challenges. First, these enzymes are highly redundant in most bacteria, where strains lacking single hydrolases usually do not show significant growth defects. Second, as the uncontrolled action of PG hydrolases potentially leads to cell lysis, it is difficult to observe highly activated PG hydrolysis during normal cell growth [37]. The sporulation process, in which vegetative *M. xanthus* cells degrade their PG thoroughly within two hours [21,22,41,106], sets the perfect stage for the study of PG hydrolases. In order to facilitate rapid PG degradation, the balance between PG polymerases and hydrolases changes, and hydrolysis becomes dominant over synthesis. Thus, studying PG hydrolases during sporulation could provide valuable insight into how PG polymerase co-ordinates with hydrolase.

In conclusion, the complete degradation of PG during chemical-induced sporulation makes *M. xanthus* an invaluable model organism for investigating the dynamics of PG and cell morphology. First, when *M. xanthus* spores germinate, cells must rebuild their walls and re-establish rod shape without pre-existing PG as a template, matching the definition of "morphogenesis" perfectly. Second, unlike spheroplasts and L-forms, nascent *M. xanthus* cells restore rod-shape within one generation [22], which largely excludes the involvement of cell division. Third, the germination progress of individual *M. xanthus* spores can be tracked using simple bright-field imaging techniques, such as phase-contrast and differential interference contrast microscopy. At the population level, germination progress can be easily quantified using the aspect ratio (length/width) of individual spores [22]. Using *M. xanthus* as a model to study PG dynamics and cell morphology may allow us to answer many questions regarding bacterial growth and survival.

Author Contributions: H.Z., S.V. and B.N. wrote the manuscript. All authors have read and agreed to the published version of the manuscript.

Funding: This research was funded by the National Institutes of Health grant number R01GM129000.

Institutional Review Board Statement: Not applicable.

Informed Consent Statement: Not applicable.

Data Availability Statement: Not applicable.

Acknowledgments: The authors thank Joshua Pettibon and Autumn McManis for critical reading of this manuscript.

Conflicts of Interest: The authors declare no conflict of interest.

References

1. Yulo, P.R.J.; Hendrickson, H.L. The evolution of spherical cell shape; progress and perspective. *Biochem. Soc. Trans.* **2019**, *47*, 1621–1634. [CrossRef]
2. Chang, F.; Huang, K.C. How and why cells grow as rods. *BMC Biol.* **2014**, *12*, 54. [CrossRef]
3. Young, K.D. The selective value of bacterial shape. *Microbiol. Mol. Biol. Rev.* **2006**, *70*, 660–703. [CrossRef]
4. Holtje, J.V. Growth of the stress-bearing and shape-maintaining murein sacculus of *Escherichia coli*. *Microbiol. Mol. Biol. Rev.* **1998**, *62*, 181–203. [CrossRef]
5. Yao, X.; Jericho, M.; Pink, D.; Beveridge, T. Thickness and elasticity of gram-negative murein sacculi measured by atomic force microscopy. *J. Bacteriol.* **1999**, *181*, 6865–6875. [CrossRef]
6. Typas, A.; Banzhaf, M.; Gross, C.A.; Vollmer, W. From the regulation of peptidoglycan synthesis to bacterial growth and morphology. *Nat. Rev. Microbiol.* **2012**, *10*, 123–136. [CrossRef] [PubMed]
7. Young, K.D. Bacterial shape: Two-dimensional questions and possibilities. *Annu. Rev. Microbiol.* **2010**, *64*, 223–240. [CrossRef] [PubMed]
8. Holtje, J.V.; Heidrich, C. Enzymology of elongation and constriction of the murein sacculus of *Escherichia coli*. *Biochimie* **2001**, *83*, 103–108. [CrossRef]
9. Egan, A.J.F.; Errington, J.; Vollmer, W. Regulation of peptidoglycan synthesis and remodelling. *Nat. Rev. Microbiol.* **2020**, *18*, 446–460. [CrossRef] [PubMed]
10. Claessen, D.; Errington, J. Cell Wall Deficiency as a Coping Strategy for Stress. *Trends Microbiol.* **2019**, *27*, 1025–1033. [CrossRef]
11. Errington, J.; Mickiewicz, K.; Kawai, Y.; Wu, L.J. L-form bacteria, chronic diseases and the origins of life. *Philos. Trans. R. Soc. B Biol. Sci.* **2016**, *371*. [CrossRef] [PubMed]
12. Ranjit, D.K.; Jorgenson, M.A.; Young, K.D. PBP1B Glycosyltransferase and Transpeptidase Activities Play Different Essential Roles during the De Novo Regeneration of Rod Morphology in *Escherichia coli*. *J. Bacteriol.* **2017**, *199*. [CrossRef]
13. Leaver, M.; Dominguez-Cuevas, P.; Coxhead, J.M.; Daniel, R.A.; Errington, J. Life without a wall or division machine in *Bacillus subtilis*. *Nature* **2009**, *457*, 849–853. [CrossRef] [PubMed]
14. Mercier, R.; Kawai, Y.; Errington, J. Excess membrane synthesis drives a primitive mode of cell proliferation. *Cell* **2013**, *152*, 997–1007. [CrossRef] [PubMed]
15. Kawai, Y.; Mickiewicz, K.; Errington, J. Lysozyme Counteracts beta-Lactam Antibiotics by Promoting the Emergence of L-Form Bacteria. *Cell* **2018**, *172*, 1038–1049. [CrossRef]
16. Kawai, Y.; Mercier, R.; Errington, J. Bacterial cell morphogenesis does not require a preexisting template structure. *Curr. Biol.* **2014**, *24*, 863–867. [CrossRef]
17. Billings, G.; Ouzounov, N.; Ursell, T.; Desmarais, S.M.; Shaevitz, J.; Gitai, Z.; Huang, K.C. De novo morphogenesis in L-forms via geometric control of cell growth. *Mol. Microbiol.* **2014**, *93*, 883–896. [CrossRef]
18. Wu, L.J.; Lee, S.; Park, S.; Eland, L.E.; Wipat, A.; Holden, S.; Errington, J. Geometric principles underlying the proliferation of a model cell system. *Nat. Commun.* **2020**, *11*, 4149. [CrossRef]
19. Muñoz-Dorado, J.; Marcos-Torres, F.J.; García-Bravo, E.; Moraleda-Muñoz, A.; Pérez, J. Myxobacteria: Moving, killing, feeding, and surviving together. *Front. Microbiol.* **2016**, *7*. [CrossRef]
20. Zusman, D.R.; Scott, A.E.; Yang, Z.; Kirby, J.R. Chemosensory pathways, motility and development in *Myxococcus xanthus*. *Nat. Rev. Microbiol.* **2007**, *5*, 862–872. [CrossRef]
21. Bui, N.K.; Gray, J.; Schwarz, H.; Schumann, P.; Blanot, D.; Vollmer, W. The peptidoglycan sacculus of *Myxococcus xanthus* has unusual structural features and is degraded during glycerol-induced myxospore development. *J. Bacteriol.* **2009**, *191*, 494–505. [CrossRef]
22. Zhang, H.; Mulholland, G.A.; Seef, S.; Zhu, S.; Liu, J.; Mignot, T.; Nan, B. Establishing rod shape from spherical, peptidoglycan-deficient bacterial spores. *Proc. Natl. Acad. Sci. USA* **2020**, *117*, 14444–14452. [CrossRef]
23. Meeske, A.J.; Riley, E.P.; Robins, W.P.; Uehara, T.; Mekalanos, J.J.; Kahne, D.; Walker, S.; Kruse, A.C.; Bernhardt, T.G.; Rudner, D.Z. SEDS proteins are a widespread family of bacterial cell wall polymerases. *Nature* **2016**, *537*, 634–638. [CrossRef] [PubMed]
24. Cava, F.; de Pedro, M.A. Peptidoglycan plasticity in bacteria: Emerging variability of the murein sacculus and their associated biological functions. *Curr. Opin. Microbiol.* **2014**, *18*, 46–53. [CrossRef] [PubMed]
25. Van Teeffelen, S.; Renner, L.D. Recent advances in understanding how rod-like bacteria stably maintain their cell shapes. *F1000Research* **2018**, *7*, 241. [CrossRef]
26. Cochrane, S.A.; Lohans, C.T. Breaking down the cell wall: Strategies for antibiotic discovery targeting bacterial transpeptidases. *Eur. J. Med. Chem.* **2020**, *194*, 112262. [CrossRef]

27. Rohs, P.D.; Buss, J.; Sim, S.I.; Squyres, G.R.; Srisuknimit, V.; Smith, M.; Cho, H.; Sjodt, M.; Kruse, A.C.; Garner, E.C. A central role for PBP2 in the activation of peptidoglycan polymerization by the bacterial cell elongation machinery. *PLoS Genet.* **2018**, *14*, e1007726. [CrossRef]
28. Sjodt, M.; Rohs, P.D.A.; Gilman, M.S.A.; Erlandson, S.C.; Zheng, S.; Green, A.G.; Brock, K.P.; Taguchi, A.; Kahne, D.; Walker, S.; et al. Structural coordination of polymerization and crosslinking by a SEDS-bPBP peptidoglycan synthase complex. *Nat. Microbiol.* **2020**, *5*, 813–820. [CrossRef]
29. Emami, K.; Guyet, A.; Kawai, Y.; Devi, J.; Wu, L.J.; Allenby, N.; Daniel, R.A.; Errington, J. RodA as the missing glycosyltransferase in *Bacillus subtilis* and antibiotic discovery for the peptidoglycan polymerase pathway. *Nat. Microbiol.* **2017**, *2*, 16253. [CrossRef] [PubMed]
30. Cho, H.; Wivagg, C.N.; Kapoor, M.; Barry, Z.; Rohs, P.D.; Suh, H.; Marto, J.A.; Garner, E.C.; Bernhardt, T.G. Bacterial cell wall biogenesis is mediated by SEDS and PBP polymerase families functioning semi-autonomously. *Nat. Microbiol.* **2016**. [CrossRef]
31. Hussain, S.; Wivagg, C.N.; Szwedziak, P.; Wong, F.; Schaefer, K.; Izore, T.; Renner, L.D.; Holmes, M.J.; Sun, Y.; Bisson-Filho, A.W.; et al. MreB filaments align along greatest principal membrane curvature to orient cell wall synthesis. *eLife* **2018**, *7*. [CrossRef]
32. Colavin, A.; Shi, H.; Huang, K.C. RodZ modulates geometric localization of the bacterial actin MreB to regulate cell shape. *Nat. Commun.* **2018**, *9*, 1280. [CrossRef]
33. Ursell, T.S.; Nguyen, J.; Monds, R.D.; Colavin, A.; Billings, G.; Ouzounov, N.; Gitai, Z.; Shaevitz, J.W.; Huang, K.C. Rod-like bacterial shape is maintained by feedback between cell curvature and cytoskeletal localization. *Proc. Natl. Acad. Sci. USA* **2014**, *111*, E1025–E1034. [CrossRef] [PubMed]
34. Pazos, M.; Vollmer, W. Regulation and function of class A Penicillin-binding proteins. *Curr. Opin. Microbiol.* **2021**, *60*, 80–87. [CrossRef]
35. Vigouroux, A.; Cordier, B.; Aristov, A.; Alvarez, L.; Ozbaykal, G.; Chaze, T.; Oldewurtel, E.R.; Matondo, M.; Cava, F.; Bikard, D.; et al. Class-A penicillin binding proteins do not contribute to cell shape but repair cell-wall defects. *eLife* **2020**, *9*. [CrossRef] [PubMed]
36. Dion, M.F.; Kapoor, M.; Sun, Y.; Wilson, S.; Ryan, J.; Vigouroux, A.; van Teeffelen, S.; Oldenbourg, R.; Garner, E.C. *Bacillus subtilis* cell diameter is determined by the opposing actions of two distinct cell wall synthetic systems. *Nat. Microbiol.* **2019**. [CrossRef] [PubMed]
37. Van Heijenoort, J. Peptidoglycan hydrolases of *Escherichia coli*. *Microbiol. Mol. Biol. Rev.* **2011**, *75*, 636–663. [CrossRef]
38. Higgins, D.; Dworkin, J. Recent progress in *Bacillus subtilis* sporulation. *FEMS Microbiol. Rev.* **2012**, *36*, 131–148. [CrossRef]
39. Tocheva, E.I.; Ortega, D.R.; Jensen, G.J. Sporulation, bacterial cell envelopes and the origin of life. *Nat. Rev. Microbiol.* **2016**, *14*, 535–542. [CrossRef]
40. McKenney, P.T.; Eichenberger, P. Dynamics of spore coat morphogenesis in *Bacillus subtilis*. *Mol. Microbiol.* **2012**, *83*, 245–260. [CrossRef]
41. Dworkin, M.; Gibson, S.M. A System for Studying Microbial Morphogenesis: Rapid Formation of Microcysts in *Myxococcus xanthus*. *Science* **1964**, *146*, 243–244. [CrossRef]
42. O'Connor, K.A.; Zusman, D.R. Starvation-independent sporulation in *Myxococcus xanthus* involves the pathway for beta-lactamase induction and provides a mechanism for competitive cell survival. *Mol. Microbiol.* **1997**, *24*, 839–850. [CrossRef] [PubMed]
43. Sudo, S.Z.; Dworkin, M. Resistance of vegetative cells and microcysts of *Myxococcus xanthus*. *J. Bacteriol.* **1969**, *98*, 883–887. [CrossRef]
44. Muller, F.D.; Schink, C.W.; Hoiczyk, E.; Cserti, E.; Higgs, P.I. Spore formation in *Myxococcus xanthus* is tied to cytoskeleton functions and polysaccharide spore coat deposition. *Mol. Microbiol.* **2012**, *83*, 486–505. [CrossRef] [PubMed]
45. Ramsey, W.S.; Dworkin, M. Microcyst germination in *Myxococcus xanthus*. *J. Bacteriol.* **1968**, *95*, 2249–2257. [CrossRef] [PubMed]
46. Elias, M.; Murillo, F.J. Induction of Germination in *Myxococcus xanthus* Fruiting Body Spores. *J. Gen. Microbiol.* **1991**, *137*, 381–388. [CrossRef]
47. Otani, M.; Inouye, M.; Inouye, S. Germination of myxospores from the fruiting bodies of *Myxococcus xanthus*. *J. Bacteriol.* **1995**, *177*, 4261–4265. [CrossRef]
48. Fu, G.; Bandaria, J.N.; Le Gall, A.V.; Fan, X.; Yildiz, A.; Mignot, T.; Zusman, D.R.; Nan, B. MotAB-like machinery drives the movement of MreB filaments during bacterial gliding motility. *Proc. Natl. Acad. Sci. USA* **2018**, *115*, 2484–2489. [CrossRef]
49. Shrestha, R.; Lockless, S.W.; Sorg, J.A. A *Clostridium difficile* alanine racemase affects spore germination and accommodates serine as a substrate. *J. Biol. Chem.* **2017**, *292*, 10735–10742. [CrossRef]
50. Hsu, Y.P.; Rittichier, J.; Kuru, E.; Yablonowski, J.; Pasciak, E.; Tekkam, S.; Hall, E.; Murphy, B.; Lee, T.K.; Garner, E.C.; et al. Full color palette of fluorescent d-amino acids for in situ labeling of bacterial cell walls. *Chem. Sci.* **2017**, *8*, 6313–6321. [CrossRef]
51. Kuru, E.; Radkov, A.; Meng, X.; Egan, A.; Alvarez, L.; Dowson, A.; Booher, G.; Breukink, E.; Roper, D.I.; Cava, F.; et al. Mechanisms of incorporation for D-amino acid probes that target peptidoglycan biosynthesis. *ACS Chem. Biol.* **2019**. [CrossRef]
52. Mueller, E.A.; Egan, A.J.; Breukink, E.; Vollmer, W.; Levin, P.A. Plasticity of *Escherichia coli* cell wall metabolism promotes fitness and antibiotic resistance across environmental conditions. *eLife* **2019**, *8*. [CrossRef]
53. Leonardy, S.; Miertzschke, M.; Bulyha, I.; Sperling, E.; Wittinghofer, A.; Sogaard-Andersen, L. Regulation of dynamic polarity switching in bacteria by a Ras-like G-protein and its cognate GAP. *EMBO J.* **2010**, *29*, 2276–2289. [CrossRef] [PubMed]
54. Zhang, Y.; Franco, M.; Ducret, A.; Mignot, T. A bacterial Ras-like small GTP-binding protein and its cognate GAP establish a dynamic spatial polarity axis to control directed motility. *PLoS Biol.* **2010**, *8*, e1000430. [CrossRef]

55. McLoon, A.L.; Wuichet, K.; Hasler, M.; Keilberg, D.; Szadkowski, D.; Sogaard-Andersen, L. MglC, a paralog of *Myxococcus xanthus* GTPase activating protein MglB, plays a divergent role in motility regulation. *J. Bacteriol.* **2015**. [CrossRef]
56. Zhang, Y.; Guzzo, M.; Ducret, A.; Li, Y.Z.; Mignot, T. A dynamic response regulator protein modulates G-protein-dependent polarity in the bacterium *Myxococcus xanthus*. *PLoS Genet.* **2012**, *8*, e1002872. [CrossRef] [PubMed]
57. Keilberg, D.; Wuichet, K.; Drescher, F.; Sogaard-Andersen, L. A response regulator interfaces between the Frz chemosensory system and the MglA/MglB GTPase/GAP module to regulate polarity in *Myxococcus xanthus*. *PLoS Genet.* **2012**, *8*, e1002951. [CrossRef]
58. Szadkowski, D.; Harms, A.; Carreira, L.A.M.; Wigbers, M.; Potapova, A.; Wuichet, K.; Keilberg, D.; Gerland, U.; Sogaard-Andersen, L. Spatial control of the GTPase MglA by localized RomR-RomX GEF and MglB GAP activities enables *Myxococcus xanthus* motility. *Nat. Microbiol.* **2019**. [CrossRef] [PubMed]
59. Pogue, C.B.; Zhou, T.; Nan, B. PlpA, a PilZ-like protein, regulates directed motility of the bacterium *Myxococcus xanthus*. *Mol. Microbiol.* **2018**, *107*, 214–228. [CrossRef]
60. Muller, F.D.; Treuner-Lange, A.; Heider, J.; Huntley, S.M.; Higgs, P.I. Global transcriptome analysis of spore formation in *Myxococcus xanthus* reveals a locus necessary for cell differentiation. *BMC Genom.* **2010**, *11*, 264. [CrossRef]
61. Munoz-Dorado, J.; Moraleda-Munoz, A.; Marcos-Torres, F.J.; Contreras-Moreno, F.J.; Martin-Cuadrado, A.B.; Schrader, J.M.; Higgs, P.I.; Perez, J. Transcriptome dynamics of the *Myxococcus xanthus* multicellular developmental program. *eLife* **2019**, *8*. [CrossRef] [PubMed]
62. Nan, B.; McBride, M.J.; Chen, J.; Zusman, D.R.; Oster, G. Bacteria that glide with helical tracks. *Curr. Biol.* **2014**, *24*, R169–R173. [CrossRef]
63. Treuner-Lange, A.; Macia, E.; Guzzo, M.; Hot, E.; Faure, L.M.; Jakobczak, B.; Espinosa, L.; Alcor, D.; Ducret, A.; Keilberg, D.; et al. The small G-protein MglA connects to the MreB actin cytoskeleton at bacterial focal adhesions. *J. Cell Biol.* **2015**, *210*, 243–256. [CrossRef] [PubMed]
64. Mauriello, E.M.; Mouhamar, F.; Nan, B.; Ducret, A.; Dai, D.; Zusman, D.R.; Mignot, T. Bacterial motility complexes require the actin-like protein, MreB and the Ras homologue, MglA. *EMBO J.* **2010**, *29*, 315–326. [CrossRef]
65. Nan, B. Bacterial Gliding Motility: Rolling Out a Consensus Model. *Curr. Biol.* **2017**, *27*, R154–R156. [CrossRef]
66. Nan, B.; Bandaria, J.N.; Guo, K.Y.; Fan, X.; Moghtaderi, A.; Yildiz, A.; Zusman, D.R. The polarity of myxobacterial gliding is regulated by direct interactions between the gliding motors and the Ras homolog MglA. *Proc. Natl. Acad. Sci. USA* **2015**, *112*, E186–E193. [CrossRef]
67. Nan, B.; Chen, J.; Neu, J.C.; Berry, R.M.; Oster, G.; Zusman, D.R. Myxobacteria gliding motility requires cytoskeleton rotation powered by proton motive force. *Proc. Natl. Acad. Sci. USA* **2011**, *108*, 2498–2503. [CrossRef]
68. Nan, B.; Bandaria, J.N.; Moghtaderi, A.; Sun, I.H.; Yildiz, A.; Zusman, D.R. Flagella stator homologs function as motors for myxobacterial gliding motility by moving in helical trajectories. *Proc. Natl. Acad. Sci. USA* **2013**, *110*, E1508–E1513. [CrossRef] [PubMed]
69. Galicia, C.; Lhospice, S.; Varela, P.F.; Trapani, S.; Zhang, W.; Navaza, J.; Herrou, J.; Mignot, T.; Cherfils, J. MglA functions as a three-state GTPase to control movement reversals of *Myxococcus xanthus*. *Nat. Commun.* **2019**, *10*, 5300. [CrossRef]
70. Faure, L.M.; Fiche, J.B.; Espinosa, L.; Ducret, A.; Anantharaman, V.; Luciano, J.; Lhospice, S.; Islam, S.T.; Treguier, J.; Sotes, M.; et al. The mechanism of force transmission at bacterial focal adhesion complexes. *Nature* **2016**, *539*, 530–535. [CrossRef]
71. Shi, H.; Bratton, B.P.; Gitai, Z.; Huang, K.C. How to Build a Bacterial Cell: MreB as the Foreman of *E. coli* Construction. *Cell* **2018**, *172*, 1294–1305. [CrossRef]
72. Reichmann, N.T.; Tavares, A.C.; Saraiva, B.M.; Jousselin, A.; Reed, P.; Pereira, A.R.; Monteiro, J.M.; Sobral, R.G.; VanNieuwenhze, M.S.; Fernandes, F.; et al. SEDS-bPBP pairs direct lateral and septal peptidoglycan synthesis in *Staphylococcus aureus*. *Nat. Microbiol.* **2019**, *4*, 1368–1377. [CrossRef]
73. Pinho, M.G.; Kjos, M.; Veening, J.W. How to get (a)round: Mechanisms controlling growth and division of coccoid bacteria. *Nat. Rev. Microbiol.* **2013**, *11*, 601–614. [CrossRef]
74. Van den Ent, F.; Izore, T.; Bharat, T.A.; Johnson, C.M.; Lowe, J. Bacterial actin MreB forms antiparallel double filaments. *eLife* **2014**, *3*, e02634. [CrossRef]
75. Shi, H.; Quint, D.A.; Grason, G.M.; Gopinathan, A.; Huang, K.C. Chiral twisting in a bacterial cytoskeletal polymer affects filament size and orientation. *Nat. Commun.* **2020**, *11*, 1408. [CrossRef] [PubMed]
76. Wong, F.; Garner, E.C.; Amir, A. Mechanics and dynamics of translocating MreB filaments on curved membranes. *eLife* **2019**, *8*. [CrossRef] [PubMed]
77. Salje, J.; van den Ent, F.; de Boer, P.; Lowe, J. Direct membrane binding by bacterial actin MreB. *Mol. Cell* **2011**, *43*, 478–487. [CrossRef] [PubMed]
78. Wingreen, N.S.; Huang, K.C. Physics of Intracellular Organization in Bacteria. *Annu. Rev. Microbiol.* **2015**, *69*, 361–379. [CrossRef] [PubMed]
79. Bratton, B.P.; Shaevitz, J.W.; Gitai, Z.; Morgenstein, R.M. MreB polymers and curvature localization are enhanced by RodZ and predict *E. coli*'s cylindrical uniformity. *Nat. Commun.* **2018**, *9*, 2797. [CrossRef]
80. Morgenstein, R.M.; Bratton, B.P.; Nguyen, J.P.; Ouzounov, N.; Shaevitz, J.W.; Gitai, Z. RodZ links MreB to cell wall synthesis to mediate MreB rotation and robust morphogenesis. *Proc. Natl. Acad. Sci. USA* **2015**. [CrossRef]

81. Kawazura, T.; Matsumoto, K.; Kojima, K.; Kato, F.; Kanai, T.; Niki, H.; Shiomi, D. Exclusion of assembled MreB by anionic phospholipids at cell poles confers cell polarity for bidirectional growth. *Mol. Microbiol.* **2017**, *104*, 472–486. [CrossRef]
82. Zielinska, A.; Savietto, A.; de Sousa Borges, A.; Martinez, D.; Berbon, M.; Roelofsen, J.R.; Hartman, A.M.; de Boer, R.; Van der Klei, I.J.; Hirsch, A.K.; et al. Flotillin-mediated membrane fluidity controls peptidoglycan synthesis and MreB movement. *eLife* **2020**, *9*. [CrossRef]
83. Garenne, D.; Libchaber, A.; Noireaux, V. Membrane molecular crowding enhances MreB polymerization to shape synthetic cells from spheres to rods. *Proc. Natl. Acad. Sci. USA* **2020**, *117*, 1902–1909. [CrossRef] [PubMed]
84. Kurita, K.; Kato, F.; Shiomi, D. Alteration of Membrane Fluidity or Phospholipid Composition Perturbs Rotation of MreB Complexes in *Escherichia coli*. *Front. Mol. Biosci.* **2020**, *7*, 582660. [CrossRef] [PubMed]
85. Hodgkin, J.; Kaiser, D. Genetics of gliding motility in *Myxococcus xanthus* (Myxobacterales): Two gene systems control movement. *Mol. Gen. Genet.* **1979**, *171*, 177–191. [CrossRef]
86. Dominguez-Escobar, J.; Chastanet, A.; Crevenna, A.H.; Fromion, V.; Wedlich-Soldner, R.; Carballido-Lopez, R. Processive movement of MreB-associated cell wall biosynthetic complexes in bacteria. *Science* **2011**, *333*, 225–228. [CrossRef]
87. Garner, E.C.; Bernard, R.; Wang, W.; Zhuang, X.; Rudner, D.Z.; Mitchison, T. Coupled, circumferential motions of the cell wall synthesis machinery and MreB filaments in *B. subtilis*. *Science* **2011**, *333*, 222–225. [CrossRef]
88. Van Teeffelen, S.; Wang, S.; Furchtgott, L.; Huang, K.C.; Wingreen, N.S.; Shaevitz, J.W.; Gitai, Z. The bacterial actin MreB rotates, and rotation depends on cell-wall assembly. *Proc. Natl. Acad. Sci. USA* **2011**, *108*, 15822–15827. [CrossRef]
89. Gitai, Z.; Dye, N.A.; Reisenauer, A.; Wachi, M.; Shapiro, L. MreB actin-mediated segregation of a specific region of a bacterial chromosome. *Cell* **2005**, *120*, 329–341. [CrossRef]
90. Oswald, F.; Varadarajan, A.; Lill, H.; Peterman, E.J.; Bollen, Y.J. MreB-Dependent Organization of the *E. coli* Cytoplasmic Membrane Controls Membrane Protein Diffusion. *Biophys. J.* **2016**, *110*, 1139–1149. [CrossRef]
91. Strahl, H.; Burmann, F.; Hamoen, L.W. The actin homologue MreB organizes the bacterial cell membrane. *Nat. Commun.* **2014**, *5*, 3442. [CrossRef]
92. Munoz-Espin, D.; Daniel, R.; Kawai, Y.; Carballido-Lopez, R.; Castilla-Llorente, V.; Errington, J.; Meijer, W.J.; Salas, M. The actin-like MreB cytoskeleton organizes viral DNA replication in bacteria. *Proc. Natl. Acad. Sci. USA* **2009**, *106*, 13347–13352. [CrossRef]
93. Madabhushi, R.; Marians, K.J. Actin homolog MreB affects chromosome segregation by regulating topoisomerase IV in *Escherichia coli*. *Mol. Cell* **2009**, *33*, 171–180. [CrossRef] [PubMed]
94. Cowles, K.N.; Gitai, Z. Surface association and the MreB cytoskeleton regulate pilus production, localization and function in *Pseudomonas aeruginosa*. *Mol. Microbiol.* **2010**, *76*, 1411–1426. [CrossRef] [PubMed]
95. Chiu, S.W.; Chen, S.Y.; Wong, H.C. Localization and expression of MreB in *Vibrio parahaemolyticus* under different stresses. *Appl. Environ. Microbiol.* **2008**, *74*, 7016–7022. [CrossRef]
96. Kruse, T.; Moller-Jensen, J.; Lobner-Olesen, A.; Gerdes, K. Dysfunctional MreB inhibits chromosome segregation in *Escherichia coli*. *EMBO J.* **2003**, *22*, 5283–5292. [CrossRef]
97. Wuichet, K.; Sogaard-Andersen, L. Evolution and diversity of the Ras superfamily of small GTPases in prokaryotes. *Genome Biol. Evol.* **2014**, *7*, 57–70. [CrossRef] [PubMed]
98. Chiou, J.G.; Balasubramanian, M.K.; Lew, D.J. Cell Polarity in Yeast. *Annu. Rev. Cell Dev. Biol.* **2017**, *33*, 77–101. [CrossRef]
99. Bonazzi, D.; Julien, J.D.; Romao, M.; Seddiki, R.; Piel, M.; Boudaoud, A.; Minc, N. Symmetry breaking in spore germination relies on an interplay between polar cap stability and spore wall mechanics. *Dev. Cell* **2014**, *28*, 534–546. [CrossRef]
100. Craig, L.; Forest, K.T.; Maier, B. Type IV pili: Dynamics, biophysics and functional consequences. *Nat. Rev. Microbiol.* **2019**, *17*, 429–440. [CrossRef]
101. Nan, B.; Zusman, D.R. Novel mechanisms power bacterial gliding motility. *Mol. Microbiol.* **2016**, *101*, 186–193. [CrossRef] [PubMed]
102. Nan, B.; Zusman, D.R. Uncovering the mystery of gliding motility in the myxobacteria. *Annu. Rev. Genet.* **2011**, *45*, 21–39. [CrossRef]
103. Wartel, M.; Ducret, A.; Thutupalli, S.; Czerwinski, F.; Le Gall, A.V.; Mauriello, E.M.; Bergam, P.; Brun, Y.V.; Shaevitz, J.; Mignot, T. A versatile class of cell surface directional motors gives rise to gliding motility and sporulation in *Myxococcus xanthus*. *PLoS Biol.* **2013**, *11*, e1001728. [CrossRef] [PubMed]
104. Wittekind, M.; Schuch, R. Cell wall hydrolases and antibiotics: Exploiting synergy to create efficacious new antimicrobial treatments. *Curr. Opin. Microbiol.* **2016**, *33*, 18–24. [CrossRef]
105. Dik, D.A.; Fisher, J.F.; Mobashery, S. Cell-Wall Recycling of the Gram-Negative Bacteria and the Nexus to Antibiotic Resistance. *Chem. Rev.* **2018**, *118*, 5952–5984. [CrossRef]
106. Dworkin, M.; Voelz, H. The formation and germination of microcysts in *Myxococcus xanthus*. *J. Gen. Microbiol.* **1962**, *28*, 81–85. [CrossRef] [PubMed]

microorganisms

Article

Quantification of *Myxococcus xanthus* Aggregation and Rippling Behaviors: Deep-Learning Transformation of Phase-Contrast into Fluorescence Microscopy Images

Jiangguo Zhang [1], Jessica A. Comstock [2], Christopher R. Cotter [1], Patrick A. Murphy [1], Weili Nie [3], Roy D. Welch [2], Ankit B. Patel [3,4] and Oleg A. Igoshin [1,5,*]

1 Department of Bioengineering, Rice University, Houston, TX 77005, USA; jiangguo.zhang@rice.edu (J.Z.); cotter@sciencesundries.com (C.R.C.); pam12@rice.edu (P.A.M.)
2 Department of Biology, Syracuse University, Syracuse, NY 13244, USA; jacomsto@syr.edu (J.A.C.); rowelch@syr.edu (R.D.W.)
3 Department of Electrical and Computer Engineering, Rice University, Houston, TX 77005, USA; monde.nie@gmail.com (W.N.); abp4@rice.edu (A.B.P.)
4 Department of Neuroscience, Baylor College of Medicine, Houston, TX 77005, USA
5 Departments of Biosciences and of Chemistry, Rice University, Houston, TX 77005, USA
* Correspondence: igoshin@rice.edu; Fax: +1-(713)-348-5877

Citation: Zhang, J.; Comstock, J.A.; Cotter, C.R.; Murphy, P.A.; Nie, W.; Welch, P.D.; Patel, A.B.; Igoshin, O.A. Quantification of *Myxococcus xanthus* Aggregation and Rippling Behaviors: Deep-Learning Transformation of Phase-Contrast into Fluorescence Microscopy Images. *Microorganisms* **2021**, *9*, 1954. https://doi.org/10.3390/microorganisms9091954

Academic Editor: David Whitworth

Received: 16 July 2021
Accepted: 31 August 2021
Published: 14 September 2021

Publisher's Note: MDPI stays neutral with regard to jurisdictional claims in published maps and institutional affiliations.

Abstract: *Myxococcus xanthus* bacteria are a model system for understanding pattern formation and collective cell behaviors. When starving, cells aggregate into fruiting bodies to form metabolically inert spores. During predation, cells self-organize into traveling cell-density waves termed ripples. Both phase-contrast and fluorescence microscopy are used to observe these patterns but each has its limitations. Phase-contrast images have higher contrast, but the resulting image intensities lose their correlation with cell density. The intensities of fluorescence microscopy images, on the other hand, are well-correlated with cell density, enabling better segmentation of aggregates and better visualization of streaming patterns in between aggregates; however, fluorescence microscopy requires the engineering of cells to express fluorescent proteins and can be phototoxic to cells. To combine the advantages of both imaging methodologies, we develop a generative adversarial network that converts phase-contrast into synthesized fluorescent images. By including an additional histogram-equalized output to the state-of-the-art pix2pixHD algorithm, our model generates accurate images of aggregates and streams, enabling the estimation of aggregate positions and sizes, but with small shifts of their boundaries. Further training on ripple patterns enables accurate estimation of the rippling wavelength. Our methods are thus applicable for many other phenotypic behaviors and pattern formation studies.

Keywords: *Myxococcus xanthus*; phase contrast microscopy; fluorescence microscopy; aggregation; rippling; deep learning; generative adversarial network

1. Introduction

Multicellular self-organization is important for biological processes across all kingdoms of life [1–5]. The development of a complicated living system requires many iterative self-organizing steps, from the creation of tissues and organs, to organisms and interacting communities. In bacteria, self-organization into biofilms can contribute to virulence in addition to providing protection from environmental stressors such as desiccation or antimicrobial agents [6]. For example, developmental self-organization is a crucial part of the myxobacterial stress response, as observed in the formation of starvation-induced fruiting bodies in a soil-dwelling bacterium *Myxococcus xanthus* [7].

M. xanthus has long been a model organism for studying collective behaviors [8]. During vegetative growth on a surface, *M. xanthus* exhibits swarming, a multicellular behavior defined by cells migrating together to efficiently colonize a substrate [9]. When a swarm encounters prey bacteria, cells organize into traveling waves of high-cell-density crests separated by lower-density troughs called ripples [10–14]. These waves may allow *M. xanthus* cells to quickly cover their prey and remain in place for longer while lysing prey cells and scavenging the resulting nutrients [14]. Nutrient limitation initiates a third self-organizing behavior, triggering the population of starving cells to aggregate into fruiting bodies where some will differentiate into environmentally resistant spores [7].

Microcinematography, i.e., time-lapse microscopy, is a useful tool for observing these behaviors during the *M. xanthus* life cycle, as time-lapse microscopic imaging is capable of capturing these dynamic biological patterns [11,13,15,16]. Phase-contrast and fluorescence microscopy are commonly used for visualizing live bacterial cells, but each comes with its set of advantages and drawbacks. While phase-contrast microscopy provides enhanced contrast for observing transparent cells on an agar substrate, it can introduce artifacts. Two common artifacts are halos at the boundary of objects [17], and shade-off [18]—gradients in intensity in large features of a specimen—which can potentially obscure aspects of the behaviors of interest. Fluorescent microscopy is useful for observing cells without the background artifacts of conventional bright-field or phase-contrast microscopy, and it provides a way to estimate the local density of bacterial populations [15]. However, the time demands of introducing genes that encode fluorescent proteins to existing mutant libraries and strain collections are often prohibitive to conducting large-scale analyses. Moreover, the additional light exposure required for fluorescent imaging can be phototoxic and, therefore, can lead to cell behavior changes and/or limit the observation duration [19].

One way to avoid the disadvantages of these imaging techniques is to develop an image-processing algorithm that transforms phase-contrast microscopy images into synthesized fluorescent images. If successful, such an algorithm could allow researchers to take advantage of the wealth of information contained in fluorescent images without the associated time constraints and complications of fluorescence microscopy. Here, we hypothesized that recent exciting advances in the field of deep learning for image processing [20,21] can be leveraged to develop and train such an algorithm. In particular, image-to-image translation has been used for tasks such as image colorization [22], image denoising [23], semantic synthesis [24], style transfer [25], etc. These applications have mostly employed Generative Adversarial Networks (GANs) [26]. The GAN model consists of two interacting neural networks, a generator and a discriminator, which are trained jointly by playing a zero-sum game. The generator tries to synthesize images to fool the discriminator, while the discriminator tries to distinguish real samples from synthesized ones. Although early GAN models suffered from low resolution [24] and a lack of detail and realistic textures [27], recently, the quality of image-to-image translation has significantly improved due to the use of more advanced conditional GAN models, such as pix2pixHD [28], which can generate high-resolution images, and SPADE [29], which can control the desired look of the output with a style image input. However, these models have never been applied to synthesize high-resolution microscopic images.

In this paper, we developed such an image-to-image transformation algorithm and assessed its performance in terms of quantification of aggregation and rippling behaviors. In particular, we prepared phase-contrast and fluorescent time-lapse movies of the aggregation and rippling phenotypes of *M. xanthus* cells labeled with tdTomato fluorescent protein, and compared the detection and segmentation of biological features in both sets of images. Based on the state-of-the-art pix2pixHD model, we then developed a new conditional GAN network architecture called pix2pixHD-HE that can be trained on phase-contrast and fluorescent images taken of the same field of view at the same time. We applied our model to transform phase-contrast images into synthesized fluorescent images. To assess the model's performance, we compared the image quality with the pix2pixHD baseline. We also compared the position and size of aggregates segmented from our synthesized images to those segmented from real fluorescent images. To determine whether the model is applicable to other self-organization patterns, we applied the model to time-lapse movies of rippling and used it to quantify the ripple wavelength. The results demonstrate the generalizability of our approach for a wide range of cell patterns, but some training could be required for new phenomena.

2. Materials and Methods

2.1. Strains and Culture Conditions

M. xanthus tdTomato-expressing strain LS3908 [15] and GFP-expressing strain DK10547 [11], and *E. coli* K12 were used in this study. *M. xanthus* strains were grown overnight at 32 °C with vigorous shaking in CTTYE broth (1% Casein Peptone (Remel, San Diego, CA, USA), 0.5% Bacto Yeast Extract (BD Biosciences, Franklin Lakes, NJ, USA), 10 mM Tris (pH 8.0), 1 mM $KH(H_2)PO_4$, 8 mM $MgSO_4$) supplemented with 10 µg/mL oxytetracycline and 1 mM isopropyl β-D-1-thiogalactopyrano-side (IPTG) for LS3908 or with 40 µg/mL kanamycin for DK10547. For development assays, mid-log phase cells were harvested, resuspended in TPM starvation buffer (10 mM Tris (pH 7.6), 1 mM $KH(H_2)PO_4$, 8 mM $MgSO_4$) to a concentration of 5×10^9 cells/mL (or 1×10^{10} cells/mL for high-density and 2.5×10^9 cells/mL for low-density samples) and plated on a microscope slide chamber prepared, as previously described [30], containing 1% agarose TPM media with 1 mM IPTG added. To track cells in streams during development, LS3908 cells were diluted 1:800 into DK10547 and plated on a microscope slide chamber as above.

To induce rippling, *E. coli* K12 cells were grown overnight in LB broth (Sigma, St. Louis, MO, USA) in a 37 °C incubator with vigorous shaking, harvested and washed in TPM buffer, and plated on 1% or 0.6% agarose microscope slide chambers containing TPM supplemented with 1 mM IPTG. Once cell spots of *E. coli* were dry, LS3908 cells from an overnight culture were prepared in TPM as above, 3 µL were plated in the center of the *E. coli* spot, and the slide was incubated in the dark at 32 °C for 8–10 h before imaging to provide time for rippling to initiate.

2.2. Time-Lapse Imaging

Microscope slide chambers were placed on a stage warmer (20/20 Technologies, Wilmington, NC, USA) set to 32 °C on a Nikon Eclipse E-400 microscope (Nikon Instruments, Melville, NY, USA). A pco.panda 4.2 sCMOS camera and NIS-Elements software were used for automated time-lapse imaging, capturing a phase contrast and fluorescent image every 60 s for a total of 24 h for development movies and 8 h for rippling movies. Phase-contrast images were taken with 70 ms exposure, and transmitted light was manually shuttered with a Uniblitz VMM-D1 shutter (Uniblitz Electronics, Rochester, NY, USA) when not actively imaging. Fluorescent tdTomato-expressing samples were imaged with 400 ms exposure with a Sola LED light source (Lumencore) at 75% intensity, and GFP-expressing samples were imaged with 200 ms exposure at 35% intensity. A MAC6000 system filter wheel controller and focus control module (Ludl Electronic Products, Ltd., Hawthorne, NY, USA) were used for control of the fluorescent filter wheel and the autofocus feature.

2.3. Image Processing

The raw images captured by phase-contrast and fluorescence microscopes have inconsistent contrast and we do not need such high-resolution images to resolve aggregates and streams. Furthermore, using original resolution for training may be too slow and require too much GPU RAM memory. Therefore, we scaled down and cropped the images to train an efficient model. The processing pipeline is shown in Figure 1. For detailed methods, see Supplementary Materials 1.1.

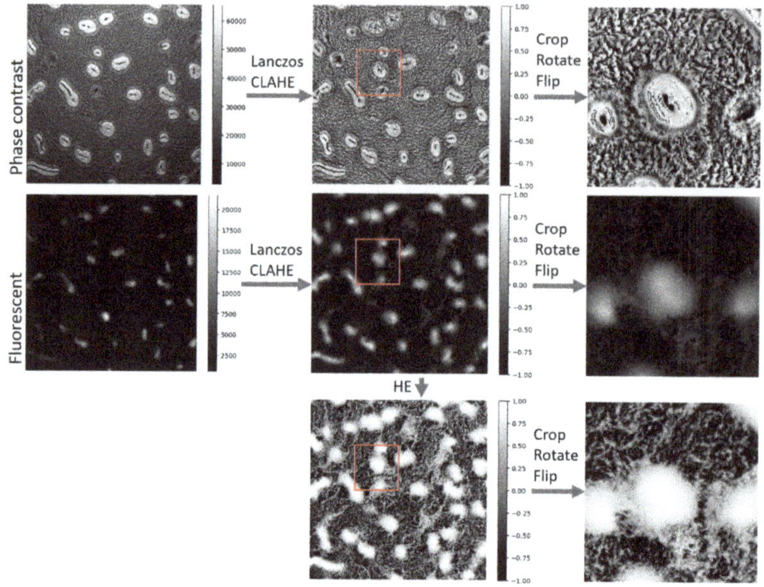

Figure 1. Image-processing pipeline. The intensity ranges are shown in the color bars, and the cropped regions are shown in red boxes. Phase contrast and fluorescent images are from the same experimental replicate and time point. LANCZOS—The Lanczos algorithm; CLAHE—Contrast Limited Adaptive Histogram Equalization.

2.4. Cell Tracking

Cell tracking was performed as described by Cotter et al. [15] for 15 min. The trajectories of tracked cells are superimposed on phase-contrast, fluorescence microscopy, and histogram-equalized fluorescence microscopy images at the initial frame of tracking.

2.5. Image Segmentation

The image segmentation algorithm converts grayscale images to binary images, where the foreground represents aggregates and the background represents the interaggregate space (Results Section 3.1). Segmentation is based on the observation that aggregates are brighter in intensity than the surrounding field and consists of two steps. First, we applied a bandpass filter as described in [15] to the input images to filter out low-frequency background and high-frequency noise. Second, we used Otsu's method [31] to perform automatic image thresholding. The algorithm returned a threshold separating the pixels into foreground and background, where the interclass difference is maximized. Finally, we applied erosion and dilation to remove small foreground spots, fill the holes in the foreground, and smooth the foreground boundaries. The detailed method is described in Supplementary Materials 1.1.4.

2.6. Network Architectures and Learning Algorithm

We adapted our generator and discriminator architectures from those in [28]. We modified the pix2pixHD framework by adding a histogram-equalized output channel in the generator (Figure 2). The pix2pixHD-HE generator network architecture is illustrated in Figure 2. We searched through different branch points in the generator architecture (Tables S1 and S2) for the best performing network architecture as measured by MSE and SSIM, while keeping other hyperparameters the same with pix2pixHD. All the networks were trained from scratch with Adam optimizer and a learning rate of 5×10^{-5} for 2000 epochs. We kept the loss weight of G2 the same and linearly decayed the loss weight of G1 for the first 500 epochs. Details of the architectural parameters and hyperparameters are provided in Supplementary Materials 1.2.

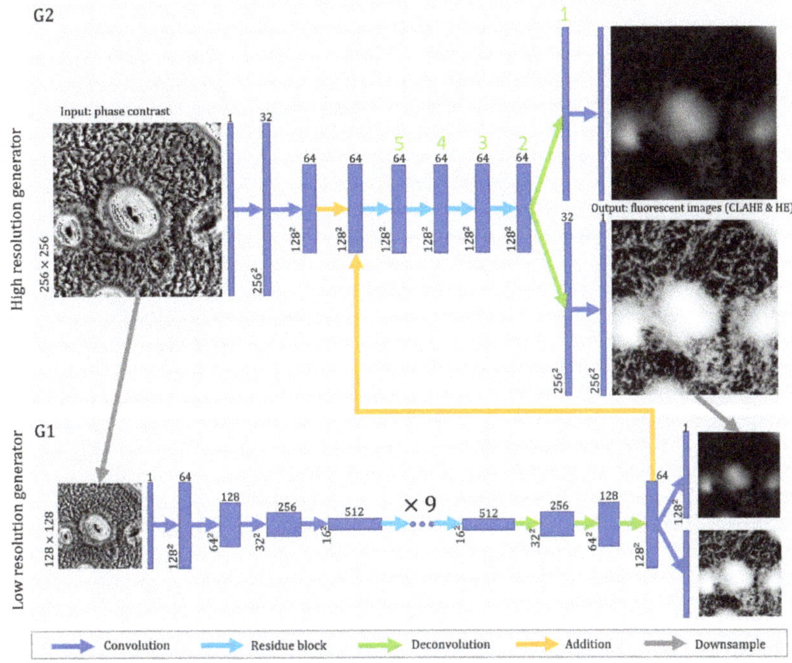

Figure 2. Network architecture of pix2pixHD-HE generator. The lengths and widths of the feature maps are annotated on the left. The depths of the feature maps are annotated as superscripts. The blue dots represent several 2-layer residual building blocks, above which the numbers of blocks are annotated. Each arrow represents an feed-forward operation step. The green numbers label the branch point, and No. 2 branch point produced the best performance (Supplementary Materials 1.2.4). We train the model as described in [28]. We use 2 discriminators to separately discriminate the normalized (CLAHE) and histogram-equalized (HE) fluorescent image channels.

2.7. Evaluation Metrics

2.7.1. Mean Square Error (MSE) and Structural Similarity Index Measure (SSIM)

To evaluate the similarity between synthesized images and real fluorescent images, we computed and compared two metrics—(i) the Mean Squared Error (MSE) and (ii) the Structural Similarity Index Measure (SSIM)—between the predicted and target images. We did not use another commonly used measure, peak signal-to-noise ratio, as it is mathematically equivalent to MSE when using normalized images. Note that unlike many image reconstruction tasks in which only high-level features need to be accurately recovered, our case requires capturing more low-level features and ideally needs to recover how cell density changes in space and time. Given that cell density is well-correlated to fluorescence microscopy image intensity [15], we need the exact match of the intensity distribution of the fluorescence image to recover biologically relevant features such as position and relative density of the aggregates and streams. Therefore, we used both MSE and SSIM as our metrics on both normalized images and histogram-equalized images as they capture pixel-to-pixel variation on different levels.

For two images represented by intensity value matrices r (real fluorescent image) and s (synthesized fluorescent image), the MSE is defined as

$$MSE(r,s) = \frac{\sum_{w}^{N_w} \sum_{l}^{N_l} (r_{l,w} - s_{l,w})^2}{N_l N_w}, \tag{1}$$

where N_l is the image length and N_w is the image width. For images with intensity scaled to the range $[-1, 1]$, the MSE is in the range $[0, 4]$. The minimal value of 0 is achieved when two images are identical for all pixels. To give the MSE a more intuitive interpretation, we applied spatial perturbations and Gaussian noise on real fluorescent images and calculated when these perturbations achieved the same MSE as between real and generated images. These comparisons are described in detail in Supplementary Materials 1.3.

The SSIM is a common alternative metric to the MSE, and captures information about differences between the luminance, contrast, and structure of two images. We did not use it to train the network, but used it solely as a comparison metric for real and synthesized images. It is defined as

$$SSIM(r,s) = \frac{2\mu_r\mu_s + c_1}{\mu_r^2 + \mu_s^2 + c_1} \frac{2\sigma_{rs} + c_2}{\sigma_r^2 + \sigma_s^2 + c_2}, \tag{2}$$

where μ_r, μ_s are the mean pixel intensities of two images, σ_r^2 and σ_s^2 are the corresponding variances, and σ_{rs} is the covariance between the two images. c_1 and c_2 are constants to avoid zero in the denominators, and the values are related to the data range. Here, $c_1 = 0.004$ and $c_2 = 0.0036$ by default for images with intensity range $[-1, 1]$. The maximum value of SSIM is 1 when two images are identical for all pixels, and the minimal value is 0.

2.7.2. Aggregate Comparison

To compare the segmented aggregates, we first matched the aggregates segmented from different images. The positions of the aggregate centroids were compared; then, the aggregates were matched based on their overlap and distance. When comparing the two images, pairs of aggregates—one from each image with the shortest distance between their centroids—are considered the same aggregate unless they have no overlapping pixels. If there is no overlapping region, or if the aggregate segmented from one image is twice as large as the other one, the aggregate pair is considered a mismatch. The distribution of distances between centroid positions and the area ratio of the matched aggregate pairs are plotted in histograms (Results Section 3.5).

Then, we compared the aggregate segmentation on a pixel level. We used precision and recall to show the segmentation accuracy, assuming segmentation of real fluorescent images to be ground truth. In that case, we labeled the segmented image with four labels (Results Section 3.5)—True positives (TP, white) are the pixels correctly identified as aggregate pixels in the generated images. False positives (FP, orange) are the pixels incorrectly identified as aggregate pixels. True negatives (TN, black) are the pixels correctly identified as interaggregate pixels. False negatives (FN, red) are the pixels incorrectly identified as interaggregate pixels. The precision is defined as the fraction of pixels in the aggregates extracted from the generated images that were correct, i.e.,

$$\text{precision} = \frac{\text{TP}}{\text{TP+FP}} \tag{3}$$

On the other hand, recall is a fraction of aggregate pixels that are correctly extracted from generated image, i.e.,

$$\text{recall} = \frac{\text{TP}}{\text{TP+FN}} \tag{4}$$

2.7.3. Rippling Wavelength Detection

To determine the ripple wavelength in an $L \times L$ rippling image, we performed a two-dimensional Discrete Fourier Transformation (DFT) in polar coordinates. To simplify the problem, we first performed a 2-D DFT in Cartesian coordinates $(x, y) \rightarrow (k_x, k_y)$

with the *scipy* package [32]; then, we transformed the absolute values to polar coordinates $(k_x, k_y) \to (k, \theta)$.

We then calculated the amplitudes for different integer-valued radii k by averaging the amplitudes across all angles θ in the radius interval $[k, k+1)$. We smoothed the amplitudes by calculating the moving average with window size 3 to remove the small fluctuations. The wave-number k_1 is determined as the peak value above a certain frequency, which was set to 6 in our 504 × 504 image. A lower-frequency corresponds to gradual changes in the image intensity; the first peak value k_0 would result in a wavelength above 84 μm, corresponding not to rippling but to large-scale features. For the majority of the images, the peak corresponds to the values of k_1. Finally, the wavelength λ_k was calculated with

$$\lambda_k = \frac{L}{k_1} \tag{5}$$

The sample rippling wavelength calculation is shown in Figure S11.

3. Results

3.1. Fluorescent Images Outperform Phase-Contrast Images in Aggregate Segmentation

We collected a total of 14 time-lapse movies taken from the onset of starvation through the first 24 h of *M. xanthus* development at varying densities (see Supplementary Materials 1.1.3 for more details). Across all replicates, aggregates formed at an average of 7.7 ± 1.2 h, meaning that the data includes preaggregation cell patterns; patterns that occur during aggregate formation; and finally, patterns seen after aggregates are stable. To compare phase-contrast and fluorescence microscopy images of aggregation, we recorded the same field of view with both methods taking simultaneous snapshots every minute over 24 h of development. In order to correct for the possibility of unequal contrast and illumination in the course of the observation, we normalized the image intensity on both channels for each frame with Contrast Limited Adaptive Histogram Equalization (CLAHE). An example image of phase contrast microscopy is shown in Figure 3A and the corresponding fluorescence microscopy image in Figure 3B. Supporting Video S1 (Supplementary Materials 2.1) shows both channels side-by-side over the course of the aggregation. Clearly, the grayscale intensity of the phase-contrast images does not directly correlate with underlying cell density. The aggregates are on average much brighter than the background but all larger aggregates contain dark bands or spots. In contrast, fluorescent images show aggregates with a more uniform intensity that are clearly brighter than the background such as in Figure 3B; moreover, intermediate intensity patterns connecting the aggregates can be seen, suggesting that these are the streams along which cells move to the aggregates or in-between the aggregates.

To further qualitatively compare two imaging modalities in their ability to extract aggregates, we adapted and fine-tuned image segmentation algorithms previously described by Cotter et al. [15] (see Methods Section 2.5 for more details). We first performed aggregate segmentation on phase-contrast images. The stable aggregate boundaries are well-defined by high-intensity pixels, and the holes in the middle are filled with our algorithm. However, for some immature (i.e., unstable) aggregates, the aggregate boundaries are not well-defined, and dark bands or spots in the middle make them difficult to distinguish from interaggregate spaces (red rectangles in Figure 3C). These aggregates are often distorted or missing when segmented from phase-contrast images. Applying the same image segmentation methodology to fluorescent images, we obtained the mature aggregates without holes. The immature aggregates are also clearly segmented (Figure 3D). Therefore, fluorescent images outperform phase-contrast images in aggregate segmentation.

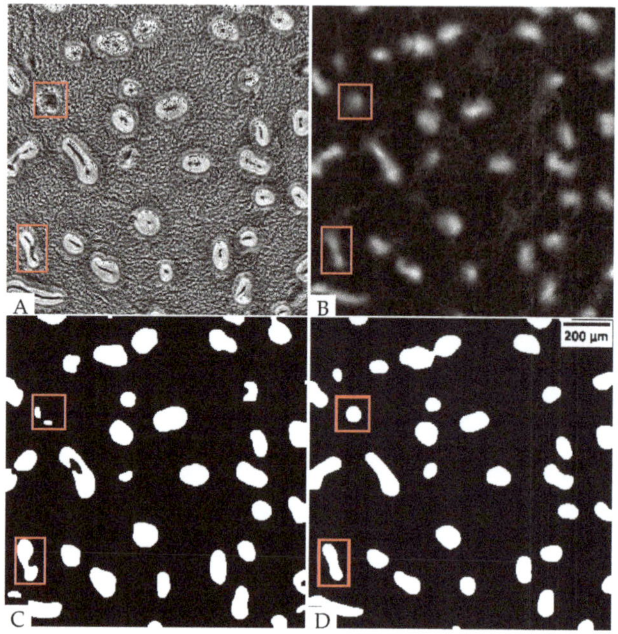

Figure 3. Aggregates extracted from different images at the 740th min: (**A**) Normalized phase-contrast image; (**B**) normalized tdTomato fluorescent image; (**C**) aggregates separated from the phase-contrast image; (**D**) aggregates separated from the fluorescent image.

3.2. Histogram Equalization (HE) Helps Visualize Streams on Fluorescent Images

To further characterize both image modalities on their ability to depict the patterns of cell motion in the interaggregate spaces, we decided to compare the patterns in image intensity with the cell trajectories. To this end, we designed an experiment to track how cells move in the stream regions by labeling a small (1:800) fraction of cells with tdTomato and the rest of the cells with GFP. We then tracked cells for 15 min (see Methods Section 2.4 for details) and superimposed these tracks onto the initial frame in each imaging modality to see if the interaggregate patterns indeed correspond to streams. As shown in Figure 4A, in the CLAHE normalized fluorescent images, cell trajectories (red) indeed somewhat align with the regions of intermediate intensity between aggregates. However, the interaggregate spacing has lower contrast than the aggregates, so the streams are not clear. To enhance the pixel contrast outside aggregates, we chose a different normalization, the so-called Histogram Equalization (HE) (Figure 4B). This normalization allows for areas of lower local contrast to gain a higher contrast. The streams on HE images are greatly enhanced and have higher intensity values than the surrounding background. As expected, most cell trajectories overlap with the high-intensity streams and the aggregates (Figure 4B).

Phase-contrast images, with (Figure 4C) or without (Figure 4D) histogram equalization, have pixel intensity values under the cell trajectories that are neither higher nor lower than the surrounding regions. This means that the cell streams are undetectable in phase-contrast images. Thus, the patterns of cell motility in the interaggregate space are also better represented in fluorescent images than in phase-contrast images.

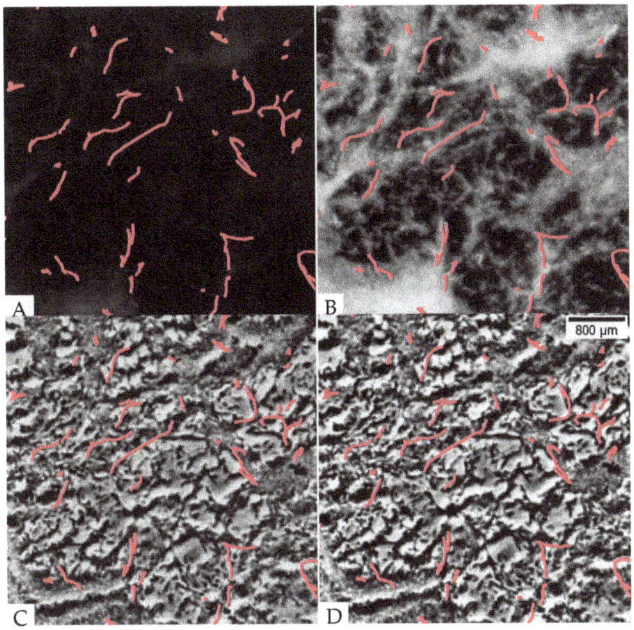

Figure 4. Fluorescent images best represent interaggregate stream patterns. Trajectories of tdTomato-expressing LS3908 cells were tracked over 15 min as they moved through a starving population of GFP-expressing DK10547 cells. Trajectories are superimposed over the corresponding (**A**) GFP fluorescent image, (**B**) histogram-equalized GFP fluorescent image, (**C**) phase-contrast image, and (**D**) histogram-equalized phase-contrast image.

3.3. Deep Neural Network Synthesizes Fluorescent Images with Aggregates and Streams

Although fluorescence microscopy is better for quantifying both the aggregates and streams in the interaggregate space, it requires engineering cells to express fluorescent proteins and may lead to cell damage due to phototoxicity [19]. To combine the advantages of both imaging methodologies, we aimed to design and train an image style transfer model to convert phase-contrast to fluorescence microscopy images. A recently developed pix2pixHD model [28] (Figure S3) has been shown to be trainable for high-resolution image transformation of semantic synthesis and sketch-to-image synthesis. We therefore first attempted to use pairs of corresponding phase-contrast and tdTomato fluorescent images as respective inputs and outputs for model training. We trained the model on 1020 images from 7 movies of 5×10^9 cells/mL and 434 images from 3 movies of 2.5×10^9 cells/mL for 2000 epochs, because a larger training set can improve the model accuracy [33], and all images in our training set have similar sizes of aggregates (Figures S9C,E and S10C,E). For validation and testing, we applied the model to images of experimental replicates with 3 different cell densities ($2.5 \times 10^9, 5 \times 10^9, 1 \times 10^{10}$ cells/mL. The results are illustrated in Figure 5 (Figures S5–S7) with synthesized images (Figures 5C and S5B–S7B) looking somewhat similar to the original fluorescent images (Figures 5B and S5A–S7A). Visually, it appears that the pix2pixHD model successfully recovers the majority of aggregates and some patterns in the interaggregate regions.

We further checked the model performance in the interaggregate space by applying histogram equalization to the synthesized image (Figures 5F and S5E–S7E) and comparing it with the histogram-equalized real fluorescent image (Figures 5E and S5D–S7D). Despite similarities in some regions, the synthesized image is blurry in areas (e.g., see the area indicated by the red box in Figure 5E) where no streaming patters are visible, or the image has uneven contrast (e.g., see the bottom region in Figures S6E and S7E). The result shows that the standard pix2pixHD model is not able to learn the low-intensity interaggregate streaming patterns.

To help the model learn streaming patterns, we changed the model architecture to include histogram-equalized fluorescent images as an extra output channel. It is necessary because histogram equalization is not an invertible process. The original image cannot be derived from the histogram-equalized image. Therefore, the resulting architecture, which we call the *pix2pixHD-HE* model (Figure 2), learns aggregate patterns from the normalized fluorescent images and interaggregate patterns from the histogram-equalized fluorescent images. We trained the model with the same training set used for the *pix2pixHD* model for 2000 epochs, and tested it on the same experimental replicate test set. The results shown in Figures 5D,G; S5C,F; S6C,F; and S7C,F demonstrate that the new model recovers both aggregate and streaming features better than pix2pixHD.

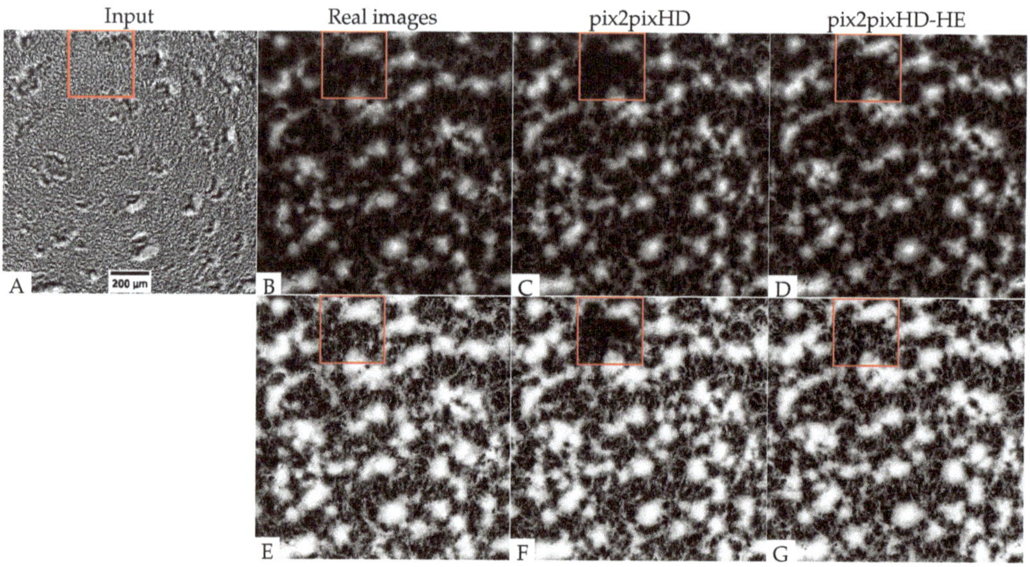

Figure 5. Model performance at the 543 min point in one training movie: (**A**) Input phase-contrast image; (**B**) ground truth tdTomato fluorescent image; (**C**) synthesized fluorescent image (pix2pixHD), the red box indicates a region where the pix2pixHD model fails to accurately reproduce the original image features; (**D**) synthesized fluorescent image (pix2pixHD-HE); (**E**) histogram-equalized ground truth tdTomato fluorescent image; (**F**) histogram-equalized synthesized fluorescent image (pix2pixHD), the red box indicates a region where the pix2pixHD model fails to accurately reproduce the original image features; (**G**) histogram-equalized synthesized fluorescent image (pix2pixHD-HE).

3.4. The Synthesized Images Show Good Global Agreement with the Real Fluorescent Images

When evaluating the synthesized image quality, we seek to get a good agreement between the image intensity and the corresponding cell density throughout the image. Therefore, we employed MSE and SSIM, image similarity evaluation measures that compare images on a pixel-by-pixel basis. The details are presented in Tables 1 and 2. We first applied our model on images of experimental replicates with cell densities of 2.5×10^9 and 5×10^9 cells/mL. The pix2pixHD-HE performs equally to pix2pixHD on both training sets and test sets, with small differences.

To check the image synthesize performance in detail, we applied the two metrics on 1440 images taken every minute in a 24-h training set, and compared the metrics for both pix2pixHD-HE and pix2pixHD across the whole movie (Figure S4A–D). From 500 to 1200 min, our pix2pixHD-HE model performed significantly better than pix2pixHD. However, outside of this time interval, the performance of pix2pixHD-HE has no significant difference from pix2pixHD. To understand why, we examined the images taken at the 500 min, 800 min, and 1200 min time points. The 500 min image corresponds to the time when the aggregates start to form—the interaggregate space is changing rapidly, and the streams are changing their location (Figure S5). By the 800 min mark, the aggregates are mostly formed and starting to mature, and the streams connecting them are forming stable spatial patterns (Figure S6). At 1200 min, most of the aggregates are mature and few cells remain in the streams. Thus, the streams connecting different aggregates are dimmer in their fluorescence intensity (Figure S7). The full dynamics of the aggregation of the real and synthesized images over 24 h is shown in Supporting Video S2 (Supplementary Materials 2.2). We conclude that pix2pixHD-HE outperforms pix2pixHD with both MSE and SSIM metrics only in the time period between 500 to 1200 min, when both aggregates and streams are present. This result suggests that the branched histogram-equalized output may assists the pix2pixHD-HE to learn the stream transformation. Notably, this time period slightly varies for different cell densities and biological replicates; nevertheless, each movie has a similar time period in which pix2pixHD-HE can better transform images than pix2pixHD.

To better quantify the difference between real and synthesized images, we calculated both a spatial displacement shift in pixels and a level of added Gaussian noise in the pixel intensity that generated equivalent differences in the values of the MSE and SSIM. For a given real image, we calculate the smallest spatial shift such that the MSE between the real and shifted image exceeds the MSE between the real and generated images. The results in pixels can be computed for every frame of the time-lapse movie; they are averaged and then converted to a single value in μm (Tables 1 and 2). A similar approach holds for the SSIM. In the same fashion, we can define a minimum standard deviation for added Gaussian noise—given as a percentage of the original intensity scale—that when added to each pixel will also give the same MSE. Again, averaging these percentages over the duration of the movie allows us to get a better interpretation of the pixel-by-pixel conversion accuracy.

We evaluated our model's performance on the images of the experimental replicates. Without histogram equalization (Table 1) on the output image, the comparable spatial shifts for the MSE are around 7.0 microns, while the relative Gaussian noise σ is around 7.8%. With SSIM, the comparable spatial differences are around 2.1 microns. After histogram equalization (Table 2), the comparable spatial differences for MSE are around 3.6 microns, while the relative Gaussian noise σ is around 19.7%. With SSIM, the comparable spatial differences are around 2.3 microns. These errors are roughly equivalent to the average cell length during development (about 7 microns [34]).

We further tested the generalizability of our model by applying it to images taken under different experimental conditions. Specifically, we tested whether our model could, without additional training, generate accurate synthesized fluorescent images of *M. xanthus* aggregation at a 2-fold-higher cell density, where the observable patterns are slightly different. Doubling the cells in the field of view leads to a proportional increase in dead cells and other debris that can impact the quality of ground truth fluorescent images (Figure S8A,B). However, these dark regions of dead cells and debris on the fluorescent images disappear as cells move over the agar to form initial aggregates, so we consider only images captured after 8 h for this analysis. The size of aggregates is generally larger in these images (Figure S8C,D), which leads to larger MSE and smaller SSIM. The comparable spatial shifts are 22.16 and 6.171 microns. This result shows that we are able to generate images for higher cell densities even with training only on images of lower cell densities.

Table 1. Model performance. SD—spatial displacement/μm; σ—minimal standard deviation for added Gaussian noise, given as a percentage of original intensity scale.

Model	Density (cells/mL)	Type	MSE Value	MSE SD	MSE σ	MSE Difference	SSIM Value	SSIM SD	SSIM Difference
pix2pixHD	5×10^9	training	0.019 ± 0.014	6.8	6.8%	-0.001 ± 0.009	0.736 ± 0.111	2.2	0.008 ± 0.032
pix2pixHD-HE	5×10^9	training	0.018 ± 0.015	6.2	6.7%		0.744 ± 0.117	2.1	
pix2pixHD	5×10^9	test	0.026 ± 0.015	6.9	8.1%	-0.002 ± 0.008	0.680 ± 0.102	2.2	0.008 ± 0.024
pix2pixHD-HE	5×10^9	test	0.025 ± 0.012	7.0	7.8%		0.687 ± 0.103	2.1	
pix2pixHD	2.5×10^9	training	0.017 ± 0.013	4.1	6.6%	0.000 ± 0.006	0.721 ± 0.128	1.9	-0.001 ± 0.032
pix2pixHD-HE	2.5×10^9	training	0.017 ± 0.011	4.3	6.6%		0.720 ± 0.133	2.0	
pix2pixHD	2.5×10^9	test	0.031 ± 0.022	5.3	8.7%	-0.001 ± 0.008	0.662 ± 0.111	2.2	0.002 ± 0.025
pix2pixHD-HE	2.5×10^9	test	0.030 ± 0.023	4.8	8.6%		0.664 ± 0.103	2.1	
pix2pixHD	1×10^{10}	test	0.119 ± 0.035	22.6	17.3%	-0.032 ± 0.007	0.420 ± 0.069	4.5	0.044 ± 0.010
pix2pixHD-HE	1×10^{10}	test	0.087 ± 0.029	16.6	14.8%		0.465 ± 0.074	4.0	

Table 2. Model performance after histogram equalization. SD—spatial displacement/μm; σ—minimal standard deviation for added Gaussian noise, given as a percentage of original intensity scale.

Model	Density (cells/mL)	Type	MSE Value	MSE SD	MSE σ	MSE Difference	SSIM Value	SSIM SD	SSIM Difference
pix2pixHD	5×10^9	training	0.101 ± 0.042	3.3	15.9%	-0.005 ± 0.010	0.458 ± 0.115	2.2	0.019 ± 0.025
pix2pixHD-HE	5×10^9	training	0.096 ± 0.043	3.2	15.5%		0.478 ± 0.123	2.2	
pix2pixHD	5×10^9	test	0.161 ± 0.074	3.7	20.1%	-0.006 ± 0.009	0.385 ± 0.125	2.4	0.016 ± 0.017
pix2pixHD-HE	5×10^9	test	0.155 ± 0.079	3.6	19.7%		0.401 ± 0.131	2.3	
pix2pixHD	2.5×10^9	training	0.098 ± 0.076	2.9	15.7%	-0.003 ± 0.006	0.487 ± 0.134	2.2	0.012 ± 0.015
pix2pixHD-HE	2.5×10^9	training	0.095 ± 0.078	2.7	15.4%		0.499 ± 0.139	2.1	
pix2pixHD	2.5×10^9	test	0.182 ± 0.064	2.5	21.3%	0.003 ± 0.010	0.338 ± 0.107	2.3	-0.001 ± 0.019
pix2pixHD-HE	2.5×10^9	test	0.184 ± 0.066	2.5	21.5%		0.336 ± 0.114	2.3	
pix2pixHD	1×10^{10}	test	0.123 ± 0.051	5.2	17.6%	-0.001 ± 0.005	0.472 ± 0.098	3.0	0.005 ± 0.008
pix2pixHD-HE	1×10^{10}	test	0.122 ± 0.049	5.1	17.5%		0.477 ± 0.094	2.9	

3.5. The Aggregates Segmented from the Synthesized Images Show Good Agreement in Their Positions and Sizes but Some Variability in Aggregate Boundaries

To evaluate the aggregate segmentation in the synthesized images, we applied the same image segmentation method on both the synthesized and the original phase-contrast images from 600 min to 1440 min, when the aggregates in three test sets were formed. We then compared both the centroid positions and sizes of corresponding aggregates between the real fluorescent images and both the synthesized and phase-contrast images. Figure 6A shows the centroid displacement from the real images for both synthesized images (blue) and phase-contrast images (orange) at a cell density of 1×10^{10} cells/mL. It is clear that the displacement of aggregates on synthesized images are smaller than that on phase-contrast images, indicating that the synthesized images more faithfully reproduce aggregate position. For images taken at a cell density of 5×10^9 cells/mL, the results are listed in Table 3. The displacement of aggregates segmented from synthesized images is greater than at high density, but it is still lower than the displacement for the lower-density phase-contrast images.

We next performed an analysis on the relative area of segmented aggregates compared with the aggregates in real fluorescent images. Figure 6B shows the relative area of aggregates in synthesized images (blue) and phase-contrast images (orange) for cell density 1×10^{10} cells/mL compared with real fluorescent images. The area of aggregates segmented from synthesized images is close to that segmented from real fluorescent images (Table 3). However, the area of aggregates segmented from phase-contrast images is smaller than those segmented from real fluorescent images, indicating that the segmentation applied to phase contrast loses parts of the aggregates. For images at a cell density of

5×10^9 cells/mL, the area of aggregates segmented from synthesized images is also close to that segmented from real fluorescent images. Again, however, the area of aggregates segmented from phase-contrast images is smaller than that segmented from real fluorescent images, albeit to a lesser extent.

The global accuracy of aggregates segmented from synthesized images was evaluated on a pixel-by-pixel basis by measuring precision (Equation (3)) and recall (Equation (4)). Precision is the fraction of correctly segmented aggregate pixels among the aggregate pixels segmented in synthesized images, whereas recall is the fraction of aggregate pixels segmented in real fluorescent images that were retrieved in synthesized images. Under both 5×10^9 and 1×10^{10} cells/mL conditions, the aggregates in synthesized images have higher precision and recall than those in phase-contrast images (Figure 6B and Table 3).

On comparing the results of synthesized images with different cell densities (Table 3), we found that our model performs better on high-cell-density images. When visualizing the aggregates extracted from high-density images (Figure 6), we found that the aggregates are larger in high-density images. As most of the distortion during synthesized image generation happens on the aggregate boundaries, larger aggregates will have a smaller displacement and relative area compared with the original. Figures S9 and S10 show the aggregate segmentation results for different cell densities and indicate that our model has higher precision and recall on large aggregates.

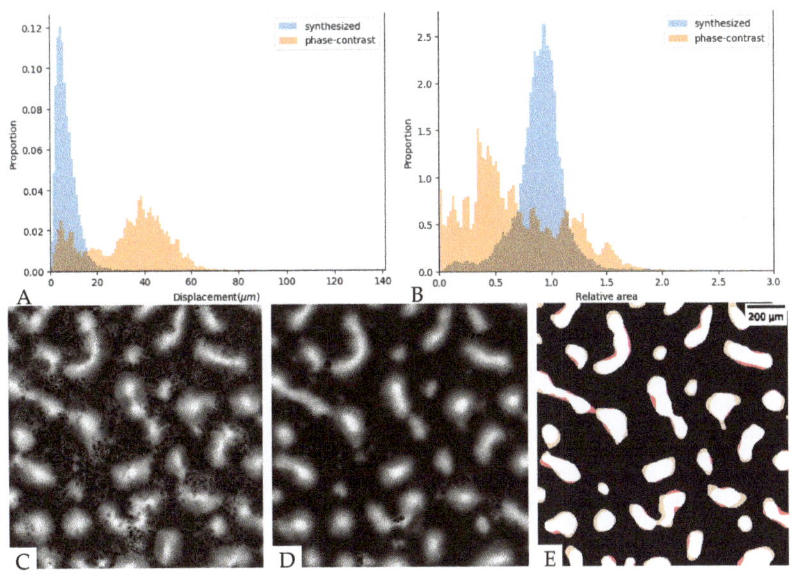

Figure 6. Aggregate segmentation for images of 1×10^{10} cells/mL cell density. (**A**) The distribution of distances between centroid positions; blue—real and synthesized fluorescent images; orange—phase-contrast and real fluorescent images. (**B**) The distribution of area ratios for the matched aggregate pairs; blue—real and synthesized fluorescent images; orange—phase-contrast and real fluorescent images. (**C**) Fluorescent image at the 800 min. (**D**) Synthesized image at the 800 min. (**E**) Aggregate segmentation result at the 800 min; white—true positive, black—true negative, yellow—false positive, red—false negative.

Table 3. Segmentation accuracy for individual aggregates.

Cell Density (cells/mL)	Image Type	Displacement (μm)	Relative Area	Precision	Recall
2.5×10^9	synthesized	4.2 ± 3.1	1.04 ± 0.31	0.82	0.88
2.5×10^9	phase-contrast	10.0 ± 7.3	1.18 ± 0.40	0.69	0.81
5×10^9	synthesized	4.4 ± 3.3	1.18 ± 0.31	0.82	0.95
5×10^9	phase-contrast	9.9 ± 10.6	1.27 ± 0.42	0.70	0.88
1×10^{10}	synthesized	7.1 ± 5.1	0.90 ± 0.24	0.90	0.83
1×10^{10}	phase-contrast	32.6 ± 16.6	0.68 ± 0.43	0.48	0.47

3.6. With Further Training, the Model Can Be Used to Convert Phase-Contrast Images of Rippling and Estimate Its Wavelength

Finally, we investigated if our model can accurately generate other self-organization patterns of *M. xanthus*. To this end, we prepared *M. xanthus* cells together with *E. coli* prey cells to induce rippling. We then recorded the same field of view with the phase-contrast and fluorescence microscopy, taking a snapshot every 60 s over 4 to 8 h while rippling occurred. When comparing the phase-contrast (Figure 7A) and fluorescence microscopy (Figure 7B) images of rippling, one can readily see that the wave-crests are much more apparent in the fluorescent images. For example, in the upper-left region of the phase-contrast image (red box in Figure 7A), the waves cannot be visibly detected, despite the clear presence of ripples in the corresponding fluorescent image. Therefore, we hypothesized that ripple wavelength would not be as accurately quantified from the phase-contrast images. To verify this, we computed the Fourier transform spectrum for each image and used the peak value above a certain threshold in the spectra to compute rippling wavelength (see Methods Section 2.7.3 for the details). The wavelength obtained from a strong peak in the spectrum for the fluorescence images ($\lambda_k \sim 46$ μm) matches the distance between neighboring crests. For phase-contrast images, the wavelength estimated from the whole image is the same ($\lambda_k \sim 46$ μm). However, in the upper-left region, the peak in the spectrum is not very strong and the wavelength corresponding to it ($\lambda_k \sim 21$ μm) is too small to match the observed ripple patterns, demonstrating that the wavelength estimation is much more accurate when quantified from fluorescent images. Thus, we test whether the rippling patterns and wavelength estimates we observe using our image transformation from phase-contrast to effective fluorescence images match those observed in the ground truth fluorescence microscopy images.

First, we apply our model trained on the aggregation patterns to rippling images without extra training. As one can see in Figure 7C, the rippling pattern is not visible in some regions of the synthesized image and the patterns are hardly more apparent in comparison to the input phase-contrast image. Not surprisingly, the wavelength estimate for that image was also low ($\lambda_k \sim 21$ μm). Next, we wanted to test if we can train the pix2pixHD-HE to learn the rippling patterns. However, it takes a significant amount of time to train a new model from scratch, and we did not have as many rippling movies as aggregate movies. We decided to use a training technique called transfer learning that can use a smaller training set and save training time.

Since we are still transforming *M. xanthus* images, just for a different experimental condition, the knowledge gained by our model to transform phase-contrast images to synthesized fluorescent images can be reused. We performed this knowledge transfer by using mapping-based deep transfer learning [35]. The previous model parameters are taken as the start point for training with the updated data set. We incorporated 731 rippling images from 2 movies in the training set and repeated training for 500 epochs. Compared with training from scratch again with 2000 epochs, this is a great increase in efficiency. This indicates that the model learned the ripple patterns based on its previous knowledge of aggregate patterns, resulting in an overall decrease in training time. The resulting model applied to Figure 7A (from a separate biological replicate, not present in the training set) shows a rippling pattern with intensity and contrast resembling that of the real fluorescent image (Figures 7B,D) over the entire field of view. The wavelength detected in the top-left region is $\lambda_k \sim 46$ µm, which exactly matches the real fluorescent image. When we calculate the wavelengths from different images across a 4-hour movie (Figure S12), the wavelengths calculated from images synthesized by the trained model are close to those calculated from real fluorescent images. On the contrary, the wavelengths in phase-contrast images are lower, and the wavelengths for the untrained model fluctuate across different frames. The result shows that our model has learned the hidden rippling features after training.

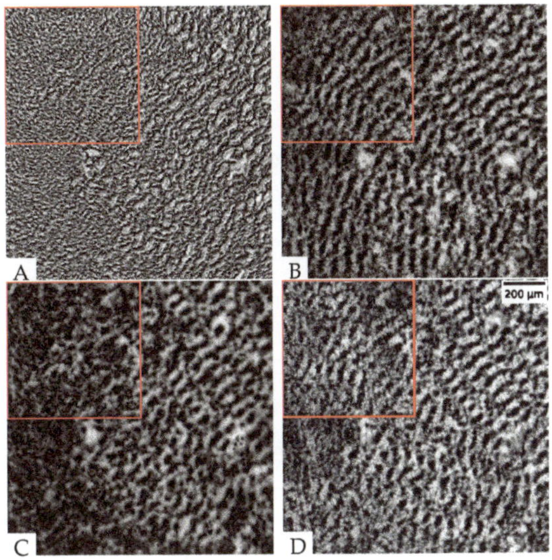

Figure 7. Rippling images and the detected wavelength λ_k in the red box. The size of the red square box is $\frac{1}{4}$ of the whole image. (**A**) Phase-contrast rippling image $\lambda_k \sim 21$ µm; (**B**) tdTomato fluorescent rippling image $\lambda_k =\sim 46$ µm; (**C**) synthesized fluorescent rippling image before training $\lambda_k \sim 63$ µm; (**D**) synthesized fluorescent rippling image after training $\lambda_k \sim 46$ µm.

4. Discussion

When choosing imaging modalities for observing live bacterial cells, it is important to consider that each one may be best at capturing a particular biological feature, but could distort or obscure other equally important features. For example, phase-contrast microscopy accentuates the contrast of nearly transparent *M. xanthus* cells to improve visualization of swarm-level patterns during aggregate development, but here, we show that the halo and shade-off effects slightly distort the image and misrepresent information about aggregate size and position (Figure 6B, Table 3) as well as the exact position of cells with respect to interaggregate streams (Figure 4). Fluorescence microscopy does not distort images in the same way, and is, therefore, better at representing aggregate features such as size and position, but at a cost; it requires a genetic mutation to introduce and express the fluorescent protein in *M. xanthus*, and an excitation light source to detect the protein in the cells. We sought to overcome the limitations of both imaging modalities by manipulating phase-contrast image data so that they could be directly compared to fluorescent image data.

To accomplish this, we built a high-resolution image transformation conditional GAN model, pix2pixHD-HE, to transform phase-contrast images to synthesized fluorescent images of *M. xanthus*. Using traditional evaluation metrics for global agreement between images—Mean Square Error (MSE) and Structural Similarity Index Measure (SSIM)—we determined that our model performs better than the state-of-the-art model pix2pixHD during aggregate formation (Figure 5). Furthermore, aggregates segmented from the converted images show excellent agreement in their position and size when compared with those segmented from real fluorescent images. Notably, with training on images of lower cell densities (2.5×10^9 and 5×10^9 cells/mL), the model can synthesize images of aggregation at high cell density (1×10^{10} cells/mL) with small error (Tables 1 and 2). With additional training, our model can also generate other multicellular dynamic patterns associated with *M. xanthus* swarms, as observed in the generation of synthesized fluorescent images of rippling that allow for accurate quantification of ripple wavelength (Figure 7D).

Fluorescence microscopes were invented to overcome the lower resolving power of ordinary optical microscopes. Although fluorescent images can better represent live-cell positions, the associated phototoxicity is an unavoidable problem that can affect cell behavior. The exposure to excitation light may change the morphology of the observed specimen [19] or could influence the behavior of bacterial cells tracked over time. Methods to reduce the impact of phototoxicity include limiting illumination to the focal plane, reducing the peak intensity, pulsing the illumination, and using longer wavelengths [19]. However, many of these methods can require specialized imaging equipment, and none of them eliminate exposure of the sample to excitation illumination altogether. Further, strains must be engineered to express the desired fluorophore, which may alter cell behavior and typically precludes high-throughput analyses involving fluorescence imaging. Our suggested approach to avoid phototoxic effects is to bypass fluorescence microscopy entirely and synthesize the fluorescent image from other nonfluorescence microscopic images, thereby eliminating excitation light exposure while retaining the ability to quantify biological patterns. Our model transforms phase-contrast images into synthesized fluorescent images, providing better resolution for observing cell behaviors while avoiding phototoxic effects.

While our model is not broadly applicable to every situation where fluorescence microscopy is required, it is particularly useful in providing a way to quantify phenotypes and behaviors that can be observed, but not accurately measured, in phase-contrast images. Our experiments with *M. xanthus* demonstrate that our model is useful in instances where an entire population of cells would be labeled with a fluorophore, as we would be able to accurately extract information about aggregate position and morphology. While phase-contrast images can approximate these data, there are times when having more accurate data on aggregate size, shape, and location is crucial, and it might be detrimental to quantify these from phase-contrast images. *M. xanthus* development is dynamic, and the factors that cause fruiting bodies to form, grow, shrink, merge, and disappear are not well understood. With training on additional data, our model could improve these investigations by synthesizing fluorescent images of wild-type and/or genetic mutant strains, without requiring the expression of a fluorophore, and quantifying aggregation patterns to determine how the mutations affect aggregation dynamics.

Additionally, in experiments where live bacterial cells are tracked, it is useful to label the subpopulation of tracked cells with one fluorophore and the rest of the cells with another so that the cell paths can be understood in the context of local cell densities and location within the swarm [15]. As shown in Figure 4C,D, cell trajectories can be superimposed over phase-contrast images, but the positions and orientations of those cells relative to streams are missing. Prior research suggests that streams may be part of fruiting body formation and growth [36]; cell behavior in streams is different than that of cells outside of streams or inside fruiting bodies [37]; and due to slime trail following [38], streams may be the avenue by which most cells enter a fruiting body. As such, our pix2pixHD-HE model could be important for improving stream visualization in cell-tracking experiments. A single fluorophore and the associated excitation illumination would still be required to track individual cells, but our model would eliminate the need for another fluorophore, and imaging of background cells could instead be done in phase-contrast and then transformed to generate a synthesized fluorescent image containing both aggregate and stream information.

In addition to the improved visualization of the features discussed above, there are some aspects of *M. xanthus* self-organizing patterns that we cannot observe with the naked eye in phase-contrast images. However, the information is still contained within these images, as our model can transform them into synthesized fluorescent images that accurately depict these patterns. For example, we found that the phase-contrast images do not show immature aggregates well, but that these aggregates are present in the real fluorescent images. (Figure 6A,C). Similarly, the rippling wavelength is sometimes undetectable with phase-contrast (Figure 7A), and when checking spatial frequencies in Fourier space, the rippling frequency is masked by the frequency of the inherent noise (Figure S11B). When we transform these phase-contrast images, the position and size of immature aggregates and the wavelength of ripples match the data extracted from the real fluorescent images (reference figures/tables). In this way, we can visualize and quantify features that were undetectable in the phase-contrast images.

The fact that the model was trained on lower cell densities and successfully recovered aggregation patterns at higher cell densities without additional training on the high-density datasets indicates that our model is likely capable of capturing the variation we see in different biological replicates under the same experimental conditions. Aggregate size and count, for example, vary from replicate to replicate, just as they vary from high to low cell densities, and we were capable of recovering them despite this variation. With additional training on rippling datasets, we were also able to use our model to recover synthesized fluorescent images of rippling from phase-contrast images without losing the ability to recover aggregates, highlighting the potential for our model to recognize and quantify multiple different biological patterns. An important caveat is that training should occur over a range of phenotypes for a specific biological pattern. Thus, if we want to use the model to transform images containing new cell patterns, it is necessary to train the model on images of the same phenotype. The specificity of the model could also be useful; however, if a synthesized image appears unrealistic, it is possible that the original image contains a pattern not present in the training set that could be biologically important but is difficult to detect just by looking at the phase-contrast images.

In this paper, we have shown an example of training a rippling image transformation model based on the aggregate image transformation model. If we want to perform a similar image transformation task for a new experimental condition, where we have fewer experimental samples, the pretrained model can be a good starting point to reuse the knowledge learned from previous training. A new model can be trained from the old using transfer learning, as we did with regards to rippling images. In other cases, we may want to extract features other than aggregates and streams. In our case, we had two outputs from pix2pixHD-HE: the normalized fluorescent images and the histogram-equalized fluorescent images. These outputs can be modified as necessary to fit new cases. For example, we could instead synthesize fluorescent images labeled with corresponding fluorophores, each used to capture different features of the bacterial dynamics. Transfer learning would allow us to use the old model wholly or in part to train a new neural network with these two new outputs. These applications of transfer learning may not only generalize our model to different *M. xanthus* phenotypes, but also to different applications within biology and medicine. The simplicity of the model training procedure indicates that it can be easily transferred to other cell systems, such as *Bacillus subtilis*, and *Escherichia coli*, bacterial model systems where self-organizing biological patterns are also important. Furthermore, medical images, such as ultrasound imaging, magnetic resonance imaging (MRI), radiographic imaging, and Cryogenic electron microscopy (Cryo-EM) imaging may also be transferred from one to another using the method detailed in this paper.

Supplementary Materials: The following are available online at https://www.mdpi.com/article/10.3390/microorganisms9091954/s1, Text S1: Supporting Methods. Figure S1: Aggregates segmented at the 800 min. Figure S2: The network architecture of pix2pixHD generator. Figure S3: The network architecture of multi-scale discriminators. Figure S4: Performance evaluation across a training movie. Figure S5: Model performance at the 500 min. Figure S6: Model performance at the 800 min. Figure S7: Model performance at the 1200 min. Figure S8: Images at the 0 and 800 min with cell density 1×10^{10} cells/mL. Figure S9: Aggregate segmentation for images of 5×10^9 cells/mL cell density. Figure S10: Aggregate segmentation for images of 2.5×10^9 cells/mL cell density. Figure S11: The cropped 504×504 rippling images and the corresponding mean amplitudes in Fourier space.

Figure S12: The calculated wavelength in a movie. Table S1: Hyperparameter optimization. Table S2: Hyperparameter optimization after histogram equalization. Video S1: Aggregate Segmentation. Video S2: Model Performance.

Author Contributions: Conceptualization, J.Z., J.A.C., A.B.P., R.D.W., and O.A.I.; methodology, J.Z., J.A.C., A.B.P., R.D.W., and O.A.I.; software, J.Z., C.R.C., W.N., and A.B.P.; validation, J.Z., A.B.P., and O.A.I.; formal analysis, C.R.C. and J.Z.; investigation, A.B.P., R.D.W., and O.A.I.; data curation, J.A.C. and R.D.W.; writing—original draft preparation, J.Z. and J.A.C.; writing—review and editing, J.Z., J.A.C., C.R.C., P.A.M., W.N., A.B.P., R.D.W., and O.A.I.; visualization, J.Z.; supervision, A.B.P., R.D.W., and O.A.I.; All authors have read and agreed to the published version of the manuscript.

Funding: This research was primarily funded by NSF IOS 1856742 and 1856665 awards with partial support of NSF IOS-1951025 and NSF DMS-1903275 awards to O.A.I and NSF DBI 1707400 award to A.B.P.

Data Availability Statement: The model is available at https://github.com/IgoshinLab/pix2pixHD-HE.git ((GitHub), accessed on 25 August 2021).

Acknowledgments: The authors thank Larry Shimkets for useful discussions at early stage of this project.

Conflicts of Interest: The authors declare no conflict of interest.

Abbreviations

The following abbreviations are used in this manuscript:

MSE	Mean Square Error
SSIM	Structural Similarity Index Measure
HE	Histogram Equalization
CLAHE	Contrast Limited Adaptive Histogram Equalization
DFT	Discrete Fourier Transformation
GAN	Generative Adversarial Network
GFP	Green Fluorescent Protein

References

1. Hartzell, T. Myxobacteria. In *eLS*; John Wiley & Sons, Ltd.: Hoboken, NJ, USA, 2016. [CrossRef]
2. Bretschneider, T.; Othmer, H.G.; Weijer, C.J. Progress and perspectives in signal transduction, actin dynamics, and movement at the cell and tissue level: Lessons from Dictyostelium. *Interface Focus* **2016**, *6*, 20160047. [CrossRef]
3. Aman, A.; Piotrowski, T. Cell migration during morphogenesis. *Dev. Biol.* **2010**, *341*, 20–33. [CrossRef]
4. Solnica-Krezel, L.; Sepich, D.S. Gastrulation: Making and shaping germ layers. *Annu. Rev. Cell Dev. Biol.* **2012**, *28*, 687–717. [CrossRef] [PubMed]
5. Theveneau, E.; Mayor, R. Neural crest delamination and migration: From epithelium-to-mesenchyme transition to collective cell migration. *Dev. Biol.* **2012**, *366*, 34–54. [CrossRef]
6. Hall-Stoodley, L.; Costerton, J.W.; Stoodley, P. Bacterial biofilms: From the natural environment to infectious diseases. *Nat. Rev. Microbiol.* **2004**, *2*, 95–108. [CrossRef] [PubMed]
7. Bretl, D.J.; Kirby, J.R. Molecular mechanisms of signaling in *Myxococcus xanthus* development. *J. Mol. Biol.* **2016**, *428*, 3805–3830. [CrossRef]
8. Zhang, Y.; Ducret, A.; Shaevitz, J.; Mignot, T. From individual cell motility to collective behaviors: Insights from a prokaryote, *Myxococcus xanthus*. *FEMS Microbiol. Rev.* **2012**, *36*, 149–164. [CrossRef]
9. Wu, Y.; Jiang, Y.; Kaiser, A.D.; Alber, M. Self-organization in bacterial swarming: Lessons from myxobacteria. *Phys. Biol.* **2011**, *8*, 055003. [CrossRef] [PubMed]
10. Shimkets, L.J.; Kaiser, D. Induction of coordinated movement of *Myxococcus xanthus* cells. *J. Bacteriol.* **1982**, *152*, 451–461. [CrossRef] [PubMed]
11. Welch, R.; Kaiser, D. Cell behavior in traveling wave patterns of myxobacteria. *Proc. Natl. Acad. Sci. USA* **2001**, *98*, 14907–14912. [CrossRef]
12. Igoshin, O.A.; Mogilner, A.; Welch, R.D.; Kaiser, D.; Oster, G. Pattern formation and traveling waves in myxobacteria: Theory and modeling. *Proc. Natl. Acad. Sci. USA* **2001**, *98*, 14913–14918. [CrossRef]
13. Berleman, J.E.; Chumley, T.; Cheung, P.; Kirby, J.R. Rippling is a predatory behavior in *Myxococcus xanthus*. *J. Bacteriol.* **2006**, *188*, 5888–5895. [CrossRef]
14. Zhang, H.; Vaksman, Z.; Litwin, D.B.; Shi, P.; Kaplan, H.B.; Igoshin, O.A. The Mechanistic Basis of *Myxococcus xanthus* Rippling Behavior and Its Physiological Role during Predation. *PLoS Comput. Biol.* **2012**, *8*, e1002715. [CrossRef] [PubMed]

15. Cotter, C.R.; Schüttler, H.B.; Igoshin, O.A.; Shimkets, L.J. Data-driven modeling reveals cell behaviors controlling self-organization during *Myxococcus xanthus* development. *Proc. Natl. Acad. Sci. USA* **2017**, *114*, E4592–E4601. [CrossRef]
16. Zhang, Z.; Cotter, C.R.; Lyu, Z.; Shimkets, L.J.; Igoshin, O.A. Data-driven models reveal mutant cell behaviors important for myxobacterial aggregation. *Msystems* **2020**, *5*, e00518-20. [CrossRef]
17. Otaki, T. Artifact halo reduction in phase contrast microscopy using apodization. *Opt. Rev.* **2000**, *7*, 119–122. [CrossRef]
18. Maurer, C.; Jesacher, A.; Bernet, S.; Ritsch-Marte, M. Phase contrast microscopy with full numerical aperture illumination. *Opt. Express* **2008**, *16*, 19821–19829. [CrossRef] [PubMed]
19. Icha, J.; Weber, M.; Waters, J.C.; Norden, C. Phototoxicity in live fluorescence microscopy, and how to avoid it. *BioEssays* **2017**, *39*, 1–15. [CrossRef]
20. Liu, M.Y.; Huang, X.; Yu, J.; Wang, T.C.; Mallya, A. Generative adversarial networks for image and video synthesis: Algorithms and applications. *Proc. IEEE* **2021**, *109*, 839–862.
21. Pang, Y.; Lin, J.; Qin, T.; Chen, Z. Image-to-Image Translation: Methods and Applications. *arXiv* **2021**, arXiv:2101.08629.
22. Suárez, P.L.; Sappa, A.D.; Vintimilla, B.X. Infrared Image Colorization Based on a Triplet DCGAN Architecture. In Proceedings of the 2017 IEEE Conference on Computer Vision and Pattern Recognition Workshops (CVPRW), Honolulu, HI, USA, 21–26 July 2017; pp. 212–217. [CrossRef]
23. Zhang, H.; Sindagi, V.; Patel, V.M. Image de-raining using a conditional generative adversarial network. *IEEE Trans. Circuits Syst. Video Technol.* **2019**, *30*, 3943–3956. [CrossRef]
24. Isola, P.; Zhu, J.Y.; Zhou, T.; Efros, A.A. Image-to-image translation with conditional adversarial networks. In Proceedings of the Proceedings—30th IEEE Conference on Computer Vision and Pattern Recognition, CVPR 2017, Honolulu, HI, USA, 21–26 July 2017. [CrossRef]
25. Zhu, J.Y.; Park, T.; Isola, P.; Efros, A.A. Unpaired Image-to-Image Translation Using Cycle-Consistent Adversarial Networks. In Proceedings of the 2017 IEEE International Conference on Computer Vision (ICCV), Venice, Italy, 22–29 October 2017; pp. 2242–2251. [CrossRef]
26. Goodfellow, I.J.; Pouget-Abadie, J.; Mirza, M.; Xu, B.; Warde-Farley, D.; Ozair, S.; Courville, A.; Bengio, Y. Generative adversarial networks. *arXiv* **2014**, arXiv:1406.2661.
27. Chen, Q.; Koltun, V. Photographic image synthesis with cascaded refinement networks. In Proceedings of the IEEE International Conference on Computer Vision, Venice, Italy, 22–29 October 2017; pp. 1511–1520.
28. Wang, T.C.; Liu, M.Y.; Zhu, J.Y.; Tao, A.; Kautz, J.; Catanzaro, B. High-Resolution Image Synthesis and Semantic Manipulation with Conditional GANs. In Proceedings of the IEEE Computer Society Conference on Computer Vision and Pattern Recognition, Salt Lake City, UT, USA, 18–22 June 2018. [CrossRef]
29. Park, T.; Liu, M.Y.; Wang, T.C.; Zhu, J.Y. Semantic image synthesis with spatially-adaptive normalization. In Proceedings of the IEEE/CVF Conference on Computer Vision and Pattern Recognition, Long Beach, CA, USA, 15–20 June 2019; pp. 2337–2346.
30. Curtis, P.D.; Taylor, R.G.; Welch, R.D.; Shimkets, L.J. Spatial Organization of *Myxococcus xanthus* during Fruiting Body Formation. *J. Bacteriol.* **2007**, *189*, 9126–9130. [CrossRef]
31. Otsu, N.; Smith, P.L.; Reid, D.B.; Environment, C.; Palo, L.; Alto, P.; Smith, P.L. A threshold selection method from gray-level histograms. *IEEE Trans. Syst. Man Cybern.* **1979**, *C*, 62–66. [CrossRef]
32. Virtanen P., Gommers R., Oliphant T.E., Haberland M., Reddy T., Cournapeau D., Burovski E., Peterson P., Weckesser W., Bright J. SciPy 1.0: fundamental algorithms for scientific computing in Python. *Nature Methods* **2020**, *17*, 261-272. [CrossRef]
33. Zhu, X.; Vondrick, C.; Fowlkes, C.C.; Ramanan, D. Do we need more training data? *Int. J. Comput. Vis.* **2016**, *119*, 76–92. [CrossRef]
34. Bhat, S.; Boynton, T.O.; Pham, D.; Shimkets, L.J. Fatty Acids from Membrane Lipids Become Incorporated into Lipid Bodies during Myxococcus xanthus Differentiation. *PLoS ONE* **2014**, *9*, 1–10. [CrossRef]
35. Tan, C.; Sun, F.; Kong, T.; Zhang, W.; Yang, C.; Liu, C. A survey on deep transfer learning. In Proceedings of the International Conference on Artificial Neural Networks, Rhodes, Greece, 4–7 October 2018; Springer: Berlin/Heidelberg, Germany 2018; pp. 270–279.
36. Kaiser, D.; Welch, R. Dynamics of fruiting body morphogenesis. *J. Bacteriol.* **2004**, *186*, 919–927. [CrossRef] [PubMed]
37. Wolgemuth, C.; Hoiczyk, E.; Kaiser, D.; Oster, G. How Myxobacteria Glide. *Curr. Biol.* **2002**, *12*, 369–377. [CrossRef]
38. Yu, R.; Kaiser, D. Gliding motility and polarized slime secretion. *Mol. Microbiol.* **2007**, *63*, 454–467. [CrossRef]

microorganisms

MDPI

Article

A Genomic Survey of Signalling in the Myxococcaceae

David E. Whitworth * and Allison Zwarycz

Institute of Biological Environmental and Rural Sciences, Aberystwyth University, Aberystwyth, SY19 7DL, UK; alz11@aber.ac.uk

* Correspondence: dew@aber.ac.uk

Received: 14 October 2020; Accepted: 3 November 2020; Published: 6 November 2020

Abstract: As prokaryotes diverge by evolution, essential 'core' genes required for conserved phenotypes are preferentially retained, while inessential 'accessory' genes are lost or diversify. We used the recently expanded number of myxobacterial genome sequences to investigate the conservation of their signalling proteins, focusing on two sister genera (*Myxococcus* and *Corallococcus*), and on a species within each genus (*Myxococcus xanthus* and *Corallococcus exiguus*). Four new *C. exiguus* genome sequences are also described here. Despite accessory genes accounting for substantial proportions of each myxobacterial genome, signalling proteins were found to be enriched in the core genome, with two-component system genes almost exclusively so. We also investigated the conservation of signalling proteins in three myxobacterial behaviours. The linear carotenogenesis pathway was entirely conserved, with no gene gain/loss observed. However, the modular fruiting body formation network was found to be evolutionarily plastic, with dispensable components in all modules (including components required for fruiting in the model myxobacterium *M. xanthus* DK1622). Quorum signalling (QS) is thought to be absent from most myxobacteria, however, they generally appear to be able to produce CAI-I (cholerae autoinducer-1), to sense other QS molecules, and to disrupt the QS of other organisms, potentially important abilities during predation of other prokaryotes.

Keywords: carotenoids; comparative genomics; development; fruiting body formation; one-component systems; quorum signalling; two-component systems; myxobacteria; Myxococcales

1. Introduction

During the evolution of new species from common ancestors, phenotypic differences often emerge as a result of lineage-specific changes in underlying signalling pathways and regulatory genes. It is therefore important to understand how signalling gene sets change as organisms evolve and to be able to relate those changes to formal taxonomies. Understanding the mutability of signalling gene sets can also provide us with insights into the ecology of contemporary organisms and the molecular mechanisms of their phenotypes.

The myxobacteria (order Myxococcales) are renowned for having exceptionally large numbers of signalling genes in their genomes [1–3]. Particularly common are serine/threonine (Ser/Thr) kinases, which regulate target proteins by reversible phosphorylation, one-component systems (OCSs), which combine a sensory domain with an 'output' response effector domain, and two-component systems (TCSs), which typically comprise a sensor histidine (auto)kinase (HK) which transfers phosphoryl groups to a partner response regulator (RR), sometimes via a phosphotransfer protein (P). Myxobacterial genomes also encode numerous transcription factors (TFs), including DNA-binding transcriptional regulators (TRs), alternative sigma factors and DNA-binding OCSs.

In 2015, just twelve myxobacterial genome sequences were publicly available (including three members of family Myxococcaceae, as currently defined), and analysis of those genomes confirmed that differences in TCS gene sets scaled with phylogenetic distances between strains [3]. TCS evolution was found to be dominated by gene gain/loss rather than point mutations or intra-gene insertions/deletions [3].

By September 2020, there were 375 myxobacterial genomes available, including 102 from the Myxococcaceal family. Although not as explosive, the last decade also saw an increase in the number of known Myxococcaceal species from 8 [4] to 24, with 14 of the new species identified as a result of genome-led taxonomy [5–7]. Family Myxococcaceae is dominated by the genera *Corallococcus* and *Myxococcus* (synonymous with *Pyxidicoccus*), which currently contain ten and twelve species, respectively [7].

The increased availability of genome sequences has allowed pan-genomic analyses of different Myxococcaceal taxa [7–9]. Such analyses have revealed small core genomes (the genes shared by every member of a taxon). For *Myxococcus xanthus*, only ~75% of genes in each genome belong to the core genome, dropping to just 9% of genes when comparing species within the *Myxococcus* genus [7]. This means that a large proportion of the genes in each myxobacterial genome belong to the accessory pan-genome (genes that are absent from some genomes). Indeed, 63% of the genes constituting the pan-genome of 11 *Myxococcus* spp. type strains were found to be unique to individual species, presumably having been acquired by lineage-specific duplication (with rapid divergence), or by horizontal transfer [7].

The acquisition of new genes by a genome carries with it the metabolic cost of reproducing those extra genes and consequently results in a fitness disadvantage. The selective advantage of being able to grow faster results in an evolutionary pressure for bacteria to streamline their genomes, rapidly losing genes that do not confer a selective advantage [10]. This pressure to streamline seems to be diminished in the myxobacteria as they possess unusually large genomes, with large accessory genomes [7,11]. It has been hypothesised that the slow growth exhibited by myxobacteria may result in the reduced pressure to streamline, in turn allowing accumulation of genes that might only occasionally provide a selective advantage [11].

We expected that assessing whether specific genes are part of the core or accessory genome would allow us to distinguish between genes which are functional and contribute to core behaviours/processes, and those which are dispensable: either awaiting loss from the genome, or only beneficial under limited circumstances, or in a subset of taxa. Due to their abundance, and the wealth of knowledge regarding their functioning, we particularly wanted to assess the patterns of conservation of myxobacterial signalling genes, to better understand which pathways and processes regulate core myxobacterial functions and which are unique to individual species or strains. We therefore investigated three exemplar myxobacterial signalling pathways: carotenogenesis, fruiting body development and quorum signalling.

Production of photo-protective carotenoid pigments is regulated by illumination in *M. xanthus* via the Car pathway. Exposure to light stimulates release of the alternative sigma factor CarQ, to direct transcription of *carS* [12–14]. CarS is an anti-repressor of the *crt* carotenoid biosynthetic genes [15–17]. The pathway is essentially a long linear signalling pathway, supplemented by a second repressor, which is directly inactivated by light [18].

When a population of myxobacteria is starved, it produces a multicellular fruiting body containing myxospores [19–22]. The regulators of fruiting body formation are dominated by TCS proteins, organised into largely discrete modules. Development requires the integration of signals from multiple modules, with the secondary messengers c-di-GMP (cyclic di-GMP) and (p)ppGpp, and two major intercellular signals: A-signal is a quorum signal, while C-signalling is a consequence of cell–cell contact [22,23].

Quorum signalling (QS) pathways are simple, and multiple independent pathways can be found in some organisms [24,25]. A synthase enzyme makes the QS signal (or autoinducer, AI), which is then typically transduced by either a DNA-binding transcription factor or a TCS pathway. Typical AIs include AI-I, AI-II, CAI-I and HAI-I (autoinducer-1, autoinducer-2, cholerae autoinducer-1 and harveyi autoinducer-1) [24,25]. Myxobacteria are not known to produce any AIs, however they have recently been shown to modulate myxobacterial behaviour [26].

The signalling pathways underpinning carotenoid production, fruiting body formation and QS are therefore very different, both in organisation and in the type of regulators involved, and we hypothesised that the pathway regulators would exhibit different patterns of conservation as a consequence. To that end, we surveyed the signalling proteins found in the genomes of four distinct groups of Myxococcaceae—the ten type strains of *Corallococcus* spp., the eleven type strains of *Myxococcus/Pyxidicoccus* spp., ten strains of *M. xanthus* and ten strains of *Corallococcus exiguus* (including four genomes described here for the first time). Our analysis included TCS, Ser/Thr kinases, sigma factors, OCS and other TRs. We did not include regulatory ncRNAs (non-coding RNAs) as they have been recently surveyed elsewhere [27].

Despite their large numbers, signalling proteins (particularly TCS proteins) were found to be enriched in the core Myxococcaceal genome. While the linear carotenogenesis pathway was wholly conserved, the conservation of components of the fruiting body network was highly variable. The Myxococcaceae also generally appear to be able to produce QS signals, and to sense/disrupt the QS of other organisms.

2. Materials and Methods

2.1. Genome Sequences

Four new genome sequences are reported here. Isolates AB016, AB031, AB051 and CA048 are all wild-type strains isolated from soils in Wales, UK [28]. Genomes were sequenced by MicrobesNG (Birmingham, UK) using Illumina Inc. (San Diego, CA, USA) Hi-Seq 2500 technology. Paired-end reads were quality-checked using BWA-MEM and assembled using Spades 3.7 and Kraken 2.0 [29–31]. Assemblies were uploaded to Genbank, wherein annotation was applied using the PGAP-4 pipeline [32].

All genome sequences and CDS (protein coding sequences) used in this study (including the four newly sequenced genomes) were subsequently downloaded from Genbank. The newly sequenced strains were identified as *C. exiguus* by calculating ANI (Average Nucleotide Identity) and dDDH (digital DNA-DNA Hybridisation) values, as described previously [9]. The four strains all gave ANI values above 95% and dDDH values above 70% when compared with the *C. exiguus* (and no other) type strain genome.

2.2. Identification of Regulatory Proteins

The P2RP webserver [33] was used to identify TRs and TCS proteins among the proteins encoded by each genome. Proteins are categorised into families by P2RP on the basis of their domain architecture, according to the scheme implemented in the P2CS and P2TF databases, as described by Ortet et al. [34,35]. Homologues of signalling proteins were identified in genomes using BLASTp (NCBI, Bethesda, MD, USA), with an e-value cut-off of 0.001, discarding hits with a percentage identity lower than 50% (30% if query sequences were non-myxobacterial), coverage less than 70% of query length and/or a bit-score lower than 50.

3. Results

3.1. Selection of Sets of Genomes

To investigate variations in signalling proteins within species and within genera, we selected ten or more genomes in each of the four taxa. Ten *M. xanthus* strains were selected, including *M. xanthus* DK1622, which is the single best-characterised myxobacterium. We selected the type strains of all 11 discrete species within the *Myxococcus* genus [7], and all ten type strains in its sister genus *Corallococcus* [6]. Finally, we selected ten isolates of *C. exiguus*, which is the most commonly isolated species within the *Corallococcus* genus [28]. If more than one genome assembly was available for a strain, the assembly with the smallest number of contigs was chosen.

Table 1. Genome properties of selected strains. A: *M. xanthus* strains, B: *Myxococcus* spp. type strains, C: *Corallococcus* spp. type strains, D: *C. exiguus* strains. In each section, the mean ± standard deviation (and as a percentage of the mean) is provided for the genome size, the % GC, and the number of protein-coding sequences. JGI = Joint Genome Institute.

A:

Genus	Species	Strain	Contigs	Size (Mbp)	% GC	CDS	Accession	Reference
Myxococcus	*xanthus*	AB022	257	9.06	68.9	6995	VHLD00000000	[11]
Myxococcus	*xanthus*	AB024B	365	9.06	68.9	7013	SRLY00000000	[11]
Myxococcus	*xanthus*	AB056	230	9.11	69.1	7065	VHLB00000000	[11]
Myxococcus	*xanthus*	CA005	227	9.11	68.9	7175	SRLV00000000	[11]
Myxococcus	*xanthus*	CA006	360	9.05	68.9	6991	SRLU00000000	[11]
Myxococcus	*xanthus*	CA010	250	9.05	68.9	6981	VHLA00000000	[11]
Myxococcus	*xanthus*	CA018	727	9.07	68.8	7102	JAAEAG000000000	[9]
Myxococcus	*xanthus*	CA023	235	9.08	68.9	7047	JAAEAH000000000	[9]
Myxococcus	*xanthus*	CA027	252	9.05	68.9	7001	WBSK00000000	[9]
Myxococcus	*xanthus*	DK1622	1	9.14	68.9	7216	GCF_000012685	[36]
				9.08 ± 0.03	68.9 ± 0.1	7059 ± 82		

B:

Genus	Species	Strain	Contigs	Size (Mbp)	% GC	CDS	Accession	Reference
Myxococcus	*xanthus/virescens*	DSM 2260	57	9.24	69.2	7340	FNAJ00000000	JGI
Myxococcus	*eversor*	AB053B	124	11.39	68.9	8751	JAAIXY010000000000	[7]
Myxococcus	*fulvus*	DSM 16525	42	10.82	70.0	8318	FOIB00000000	JGI
Myxococcus	*hansupus*	Mixupus	1	9.49	69.2	7069	GCA_000280925	[5]
Myxococcus	*llanfairPGensis*	AM401	1077	12.41	68.7	9508	VIFM00000000	[7]
Myxococcus	*macrosporus*	DSM 14697	1	8.97	70.6	6966	GCA_002305895	[37]
Myxococcus	*stipitatus*	DSM 14675	1	10.35	69.2	7796	GCA_000331735	[38]
Myxococcus	*vastator*	AM301	1008	8.99	69.9	7055	JAAIYB000000000	[7]
Pyxidicoccus	*caerfyrddinensis*	CA032A	177	12.67	70.2	9986	JAAIYA000000000	[7]
Pyxidicoccus	*fallax*	DSM 14698	825	13.53	70.5	10513	JABBJJ000000000	[7]
Pyxidicoccus	*trucidator*	CA060A	136	12.67	70.3	9355	JAAIXZ000000000	[7]
				10.96 ± 1.68	69.7 ± 0.6	8423 ± 1279		

Table 1. *Cont.*

C:								
Genus	**Species**	**Strain**	**Contigs**	**Size (Mbp)**	**% GC**	**CDS**	**Accession**	**Reference**
Corallococcus	*aberyswythensis*	AB050A	625	9.98	70.0	7905	RAWK00000000	[8]
Corallococcus	*carmarthensis*	CA043D	530	10.79	69.9	8511	RAWE00000000	[8]
Corallococcus	*coralloides*	DSM 2259	1	10.08	69.9	7893	GCA_000255295	[39]
Corallococcus	*exercitus*	AB043A	961	10.15	70.3	8018	RAVW00000000	[8]
Corallococcus	*interemptor*	AB047A	459	9.47	70.0	7566	RAWM00000000	[8]
Corallococcus	*llansteffanensis*	CA051B	1244	10.53	70.3	8137	RAWB00000000	[8]
Corallococcus	*praedator*	CA031B	1491	10.51	69.7	8167	RAWI00000000	[8]
Corallococcus	*sicarius*	CA040B	802	10.39	70.2	7877	RAWG00000000	[8]
Corallococcus	*terminator*	CA054A	863	10.35	69.5	8008	RAVZ00000000	[8]
Corallococcus	*exiguus*	DSM 14696	36	10.41	69.6	8112	JAAAPK000000000	[8]
				10.27 ± 0.37	69.9 ± 0.3	8019 ± 245		

D:								
Genus	**Species**	**Strain**	**Contigs**	**Size (Mbp)**	**% GC**	**CDS**	**Accession**	**Reference**
Corallococcus	*exiguus*	AB004	735	10.60	69.4	8223	RAWS00000000	[8]
Corallococcus	*exiguus*	AB016	1212	10.75	69.6	8940	JABEKY000000000	This Study
Corallococcus	*exiguus*	AB018	647	10.45	69.4	8185	RAWR00000000	[8]
Corallococcus	*exiguus*	AB030	552	10.63	69.6	8334	RAWQ00000000	[8]
Corallococcus	*exiguus*	AB031	611	10.43	69.7	8356	JABEKZ000000000	This Study
Corallococcus	*exiguus*	AB032C	298	10.45	69.5	8078	RAWP00000000	[8]
Corallococcus	*exiguus*	AB038B	471	10.77	69.3	8409	RAWO00000000	[8]
Corallococcus	*exiguus*	AB051	1378	10.76	69.6	9025	JABELA000000000	This Study
Corallococcus	*exiguus*	CA041A	794	10.26	69.5	8071	RAWF00000000	[8]
Corallococcus	*exiguus*	CA048	723	10.35	69.6	8428	JABELB000000000	This Study
				10.55 ± 0.18	69.5 ± 0.1	8405 ± 330		

The strains selected, their taxonomy and the characteristics of their genome sequences are presented in Table 1. As would be expected, genome metrics are more variable amongst the type strains within a genus than among strains from within a single species (with the exception of *C. exiguus* strains, which have an unusually variable number of CDS). All strains possess typical myxobacterial genomes: large (9–13.5 Mbp), with high % GC contents (69–71%).

3.2. MyxococcacealGenomes Encode Similar Numbers and Types of Regulatory Proteins

Regulatory proteins were then identified among the genome-encoded CDSs for each genome (Table 2, Supplementary Table S1). TCS proteins, OCSs, TRs and alternative sigma factors were identified and categorised using P2RP [33], while Ser/Thr Kinases were identified using BLASTp, queried with Pkn8 and Pkn14 from *M. xanthus* (MXAN_1710 and MXAN_5116), which both contain the pfam domain Pkinase, PF00069 [40]. To compare the variability in numbers of the different types of proteins between groups of genomes, Table 2 also presents the variability coefficient (standard deviation divided by mean) for each type of protein in each group of genomes, expressed as a percentage.

The numbers of TCS proteins and Ser/Thr kinases identified closely match those few published previously [1–3], with most Myxococcaceal genomes encoding around 300 TCS proteins, 100 Ser/Thr kinases and 300 transcription factors. Within each set of genomes, the numbers of each type of signalling protein are broadly similar. For instance, among the set of ten *M. xanthus* genomes, the variability coefficient was less than 10% for every type of protein except for TCS phosphotransfer proteins, which are typically present in very small numbers. For TCS and RRs, the variability coefficient was particularly low: less than 1%, compared to a variability coefficient of 1.16% for the number of CDS. A similar pattern of variability was seen within the set of ten *C. exiguus* genomes, with the number of encoded phosphotransfer proteins being highly variable, but with minimal variations in the numbers of HKs and RRs (Table 2). At a genus level, more variability was seen in the numbers of all classes of regulatory genes than when considering sets of strains within a species, with *Myxococcus/Pyxidicoccus* spp. genomes exhibiting more variability in numbers of regulatory genes than those of *Corallococcus* spp. It is also noteworthy that the average *M. xanthus* genome encodes substantially fewer regulatory proteins of every type than typical for *Myxococcus/Pyxidicoccus* spp. (Table 2), and that OCS numbers were particularly variable in each taxon.

Table 2. Variability in the numbers of regulatory genes per genome. For each class of protein, the mean number (± standard deviation (sd)), and the variability coefficient (standard deviation as a function of the mean) are presented, for four taxonomic groupings (the number of strains in each taxonomic grouping is indicated in parentheses). Variability coefficients greater than 10% are in **bold**, and values for genome-wide numbers of CDS (protein coding sequences) are provided for comparison. Gray rows represent classes, while white rows are sub-classes of the class above. TCS (two-component system) and TF (transcription factor) proteins are subdivided into different sub-classes based on domain organisation. HK = histidine kinase, P = phosphotransfer protein, RR = response regulator, TR = transcriptional regulator, OCS = one-component system.

	M. xanthus **Strains (10)**		***Myxococcus/Pyxidicoccus*** **spp. (11)**		***Corallococcus*** **spp. (10)**		***C. exiguus*** **Strains (10)**	
	(mean ± sd)	**(% Variability)**	**(mean ± sd)**	**(% Variability)**	**(mean ± sd)**	**(% Variability)**	**(mean ± sd)**	**(% Variability)**
TCS	280 ± 2	0.81	329 ± 51	**15.58**	306 ± 12	4.05	302 ± 3	1.15
HK	141 ± 2	1.62	174 ± 35	**19.94**	160 ± 6	3.98	154 ± 2	1.00
P	3 ± 1	**22.22**	3 ± 1	**33.92**	3 ± 1	**31.43**	4 ± 2	**40.9**
RR	136 ± 1	0.90	151 ± 17	**11.13**	143 ± 7	4.56	143 ± 2	1.05
S/T Kinases	92 ± 5	5.25	124 ± 31	**25.00**	107 ± 8	7.00	113 ± 5	4.43
TF	263 ± 5	1.73	363 ± 79	**21.86**	341 ± 24	7.04	358 ± 10	2.89
TR	123 ± 2	1.33	171 ± 38	**22.04**	165 ± 11	6.92	178 ± 10	5.52
OCS	36 ± 3	7.18	70 ± 27	**38.46**	73 ± 9	**11.58**	71 ± 7	9.71
RR TF	50 ± 0	0.85	58 ± 8	**13.42**	48 ± 4	8.43	53 ± 1	1.96
σ factors	55 ± 1	1.77	65 ± 12	**19.08**	55 ± 3	5.75	56 ± 2	4.04
CDS	7059 ± 82	1.16	8423 ± 1279	**15.18**	8019 ± 245	3.06	8405 ± 330	3.93

3.3. *Different Families of Regulators Exhibit Distinct Patterns of Conservation*

TCS and TF proteins were sub-categorised into families on the basis of domain organisation according to the P2RP scheme, and the results are provided in Supplementary Table S1 [7]. Figure 1 shows the profile of RR protein families for all 41 genomes, while Table 3 provides the numbers of each family for selected protein families in each genome. The numbers of proteins in each family are broadly similar across all genomes, however, there are some consistent differences between and within groups of genomes. As noted above, for different protein classes, greater variability is observed when comparing protein families within a genus rather than within a species.

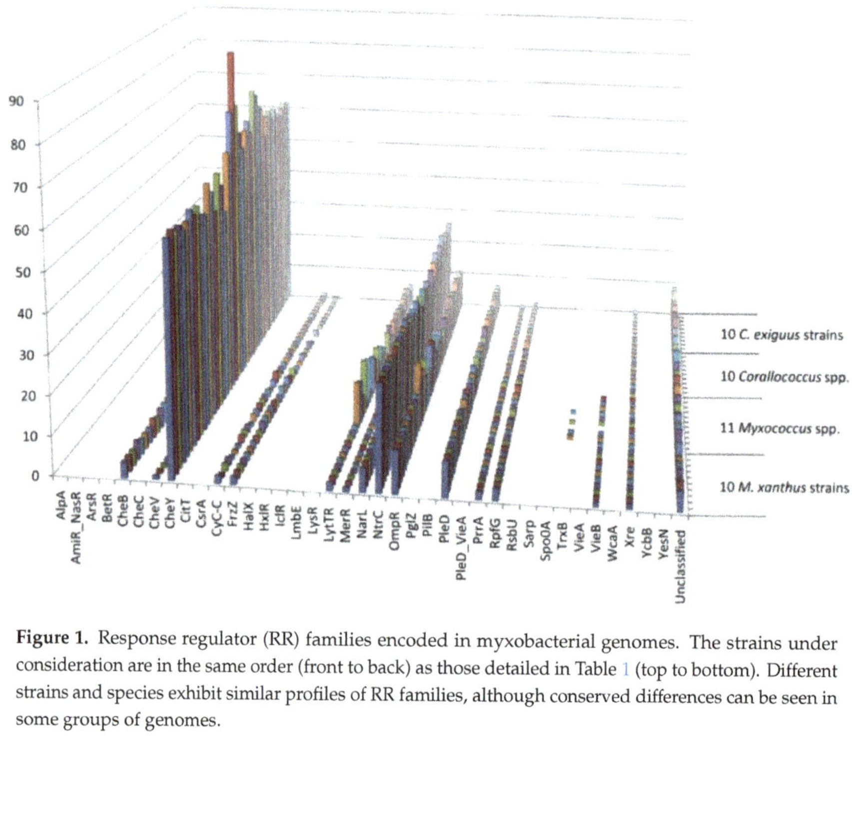

Figure 1. Response regulator (RR) families encoded in myxobacterial genomes. The strains under consideration are in the same order (front to back) as those detailed in Table 1 (top to bottom). Different strains and species exhibit similar profiles of RR families, although conserved differences can be seen in some groups of genomes.

Table 3. Numbers of illustrative protein family members per genome. Genomes are presented in four taxonomic groupings with the same background shade. Also shown are the inferred numbers of proteins in the core and accessory genomes of *M. xanthus* strains.

Protein Class	P	P	RR	RR	RR	RR	RR	RR	RR	RR	TR	TR	TR	TR	TR	TR
Protein Family	**HisKA**	**Hpt**	**CheV**	**CyC-C**	**LytTR**	**OmpR**	**PrrA**	**TrxB**	**VieB**	**Xre**	**MerR**	**Fur**	**HrcA**	**NrdR**	**PucR**	**Rok**
M. xanthus AB022	3		1	2	2	11	2		1	1	5	2	1	1		1
M. xanthus AB024B	3		1	1	2	11	2		1	1	5	2	1	1		1
M. xanthus AB056	2		1	2	2	11	2		1	1	5	2	1	1		1
M. xanthus CA005	4		1	2	2	11	2		1	1	4	2	1	1		1
M. xanthus CA006	3		1	1	2	11	2		1	1	5	2	1	1		1
M. xanthus CA010	3		1	1	2	11	2		1	1	5	2	1	1		1
M. xanthus CA018	2		1	1	2	11	2		1	1	6	2	1	1		1
M. xanthus CA023	3		1	2	2	11	2		1	1	5	2	1	1		1
M. xanthus CA027	3		1	2	2	11	2		1	1	5	2	1	1		1
M. xanthus DK1622	4		1	2	2	11	2		1	1	4	2	1	1		1
Core (per *M. xanthus* strain)	2	0	1	1	2	11	2	0	1	1	4	2	1	1	0	1
Accessory (per *M. xanthus* strain)	1	0	0	0.6	0	0	0	0	0	0	0.9	0	0	0	0	0
M. xanthus (virescens) DSM 2260	2		1	1	2	11	2		1	1	5	2	1	1		1
M. eversor AB053B	3		1	2	12	18	2	1	1	1	6	2	1	1	1	1
M. fulvus DSM 16525	3		1	2	11	13	2	1	1	1	7	3	1	1		1
M. hansupus Mixupus	2		1	2	6	14	2			1	4	2	1	1		1
M. llanfairPGensis AM401	5		1	2	14	15	2	1	1	1	7	2	1	1		1
M. macrosporus DSM 14697	3		1	2	2	10	2		1	1	5	2	1	1		1
M. stipitatus DSM 14675	3		1	2	12	13	2	1	1	1	5	2	1	1		
M. vastator AM301	5		1	1	2	9	2		1	1	5	2	1	1		1
P. caerfyrddinensis CA032A	3		1	2	10	14	2		1	1	7	2	1	1		1
P. fallax DSM 14698	2	1	1	3	8	14	2		1	1	5	2	1	1	1	1
P. trucidator CA060A	3		1	2	10	13	2			1	7	2	1	1	1	1

Table 3. *Cont.*

Protein Class	P	P	RR	RR	RR	RR	RR	RR	RR	RR	TR	TR	TR	TR	TR	TR
Protein Family	**HisKA**	**Hpt**	**CheV**	**CyC-C**	**LytTR**	**OmpR**	**PrrA**	**TrxB**	**VieB**	**Xre**	**MerR**	**Fur**	**HrcA**	**NrdR**	**PucR**	**Rok**
C. aberyswythensis AB050A	3		1	2	5	9	1			1	6	2	1	1	1	1
C. carmarthensis CA043D	3		1	2	8	12	1			1	6	2	1	1	1	1
C. coralloides DSM 2259	2		1	2	5	9	1			1	7	2	1	1	1	1
C. exercitus AB043A	3		1	2	5	11	1			1	8	2	1	1		1
C. interemptor AB047A	2		1	2	2	8	1			1	5	2	1	1		1
C. llansteffanensis CA051B	4		1	2	7	9	1			1	5	2	1	1		1
C. praedator CA031B	5		1	2	8	11	1			1	7	2	1	1		1
C. sicarius CA040B	3		1	2	4	10	1			1	8	2	1	1		1
C. terminator CA054A	3		1	2	7	10	1			1	6	2	1	1		1
C. exiguus DSM 14696	2		1	2	6	10	1			1	6	2	1		1	1
C. exiguus AB004	3		1	2	7	13	1			1	6	2	1	1	1	1
C. exiguus AB016	8		1	2	7	13	1			1	6	2	1	1	1	1
C. exiguus AB018	3	1	1	2	6	11	1			1	5	2	1	1	1	1
C. exiguus AB030	3	1	1	2	6	11	1			1	5	2	1	1	1	1
C. exiguus AB031	3		1	2	7	13	1			1	6	2	1	1	1	1
C. exiguus AB032C	3		1	2	7	13	1			1	7	2	1	1	1	1
C. exiguus AB038B	3		1	2	8	12	1			1	6	2	1	1	1	1
C. exiguus AB051	5		1	2	7	13	1			1	6	2	1	1	1	1
C. exiguus CA041A	2	1	1	2	5	11	1			1	6	2	1	1	1	1
C. exiguus CA048	3		1	2	6	11	1			1	6	2	1	1	1	1

Some protein families, for example Hpt proteins, are present in small numbers, in some but not all members of a group of genomes (a single Hpt protein each is found in the genomes of just *P. fallax* and three strains of *C. exiguus*). Such proteins are not components of the core genome and are most likely to have been acquired recently by horizontal gene transfer. Other examples include the TrxB response regulator, which is found in four of the eleven *Myxococcus/Pyxidicoccus* spp. genomes, and HisKA phosphotransfer proteins, which, when present, are found in small and highly variable numbers.

Some other protein families are found in small numbers in each genome (or each genome within a group), but at a constant number. These proteins are therefore part of the core genome, and illustrative examples include CheV, HrcA, NrdR, Rok and Xre (one in each genome), VieB (one in each *Myxococcus/Pyxidicoccus* genome, but absent from *Corallococcus* genomes) and PucR (one in each *C. exiguus* genome, but only present sporadically in *Myxococcus/Pyxidicoccus* spp. and *Corallococcus* spp. genomes). Single PrrA family members are found consistently in *Corallococcus* genomes, with two members in each *Myxococcus/Pyxidicoccus* genome, Fur consistently has two members in every genome, while Cyc-C has two members per *Corallococcus* genome but one to two in *Myxococcus/Pyxidicoccus* genomes. There are two LytTR members encoded in each *M. xanthus* genome, but highly variable numbers among *Myxococcus/Pyxidicoccus* spp. members (from two to fourteen), suggesting that two of the LytTR members are part of the core *Myxococcus/Pyxidicoccus* genome, and any others are in the accessory genome.

Many protein families have larger numbers of members in each genome, and the numbers can be highly variable (for example the MerR family of TRs and OCSs has 4–6 members in each *M. xanthus* genome), or remarkably consistent (for example the OmpR family has exactly eleven members in each *M. xanthus* genome). Presumably, for each of these larger families, there will be a core set of proteins found in each genome, and a variable number of proteins from the accessory gene pool. We would therefore consider all eleven OmpR members to be core, and four MerR members to be core, with the other MerR members being part of the accessory genome.

3.4. TCS Proteins are More Enriched in Myxobacterial Core Genomes than Other Regulatory Proteins

To investigate the relative distribution of proteins between the core and accessory genome, we categorised the proteins in each family as 'core' or 'accessory' for each group of genomes. For this purpose, we defined the number of core proteins as simply the mean number of family members, minus one standard deviation (rounded to the closest integer), with the remainder of the proteins being categorised as members of the accessory genome. The results of such an approach are provided within Table 3 for the illustrative protein families therein, for the ten *M. xanthus* genomes. The results of this simple categorisation agree well with an intuitive assessment of core vs. accessory genome membership (Table 3).

Taking this approach, and summing the results for each protein family, we were able to compare the tendencies of RRs and TFs/OCSs to be found in the core or the accessory genome of each group of genomes (Figure 2). As expected, the percentage of proteins in the core of the pan-genome is less for *Myxococcus* spp. than for *M. xanthus* strains, as the former have more diverse genomes (similarly when comparing *Corallococcus* spp. with *C. exiguus* strains), and the *Myxococcus* spp. genomes had a smaller core than *Corallococcus* spp., reflecting their greater diversity and lower percentage core genome, as described previously [7]. Similarly, a greater proportion of *C. exiguus* regulators were found to be accessory, compared with those of *M. xanthus*, which agrees with the greater variability in the numbers of regulators in their genomes, as seen in Table 1.

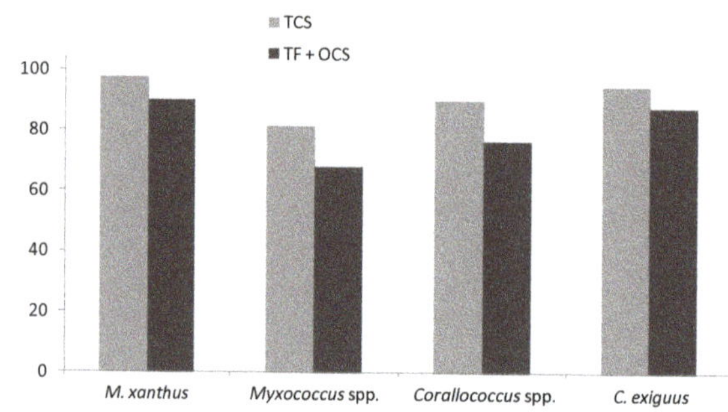

Figure 2. The percentage of RRs and TFs + OCSs found in the pan-genome core for the four groups of genomes.

In all four groups of genomes, TCS proteins were assigned to the core to a greater extent than TFs/OCSs (Figure 2), suggesting that accessory TCS proteins acquired by 'recent' horizontal gene transfer are purged from the genome faster than accessory TFs/OCSs. Possibly because recently acquired TCS proteins have the potential to disrupt pre-existing core TCS networks, while the expression of recently acquired TFs/OCS might be less likely to affect the functioning of core TFs/OCSs.

3.5. *Conservation of Regulatory Proteins Involved in Key Myxococcaceal Behaviours*

To further investigate the evolution of regulatory networks in Myxococcaceal genomes, we assessed the conservation of regulatory proteins in three 'case studies' of myxobacterial behaviours: carotenoid synthesis, fruiting body formation and quorum sensing. The regulatory mechanisms underpinning each of these phenomena are well-described and involve different classes of regulatory proteins. Identification of homologues was undertaken using BLAST, using the *M. xanthus* DK1622 protein as a query sequence (Supplementary Table S2).

Supplementary Table S2 also shows the pattern of conservation of regulators involved in the three behaviours. Regulatory proteins were designated as 'absent' from a group of n genomes if no homologues were identified in at least n-1 genomes. If the same number of homologues were found in at least n-1 genomes, the protein was denoted 'constant', and if the numbers of homologues were different in at least two genomes, the protein was classified as 'variable'. Regulators were then classified as 'core' (if homologues were found to be present at a constant number in all groups of genomes), 'conserved' (if present but found in different numbers within groups or in different groups of genomes), or 'accessory' (if absent from at least one group of genomes, or at least two genomes within a 'variable' group of genomes). Figure 3 shows the pattern of conservation of regulatory proteins involved in carotenogenesis, fruiting body formation and QS.

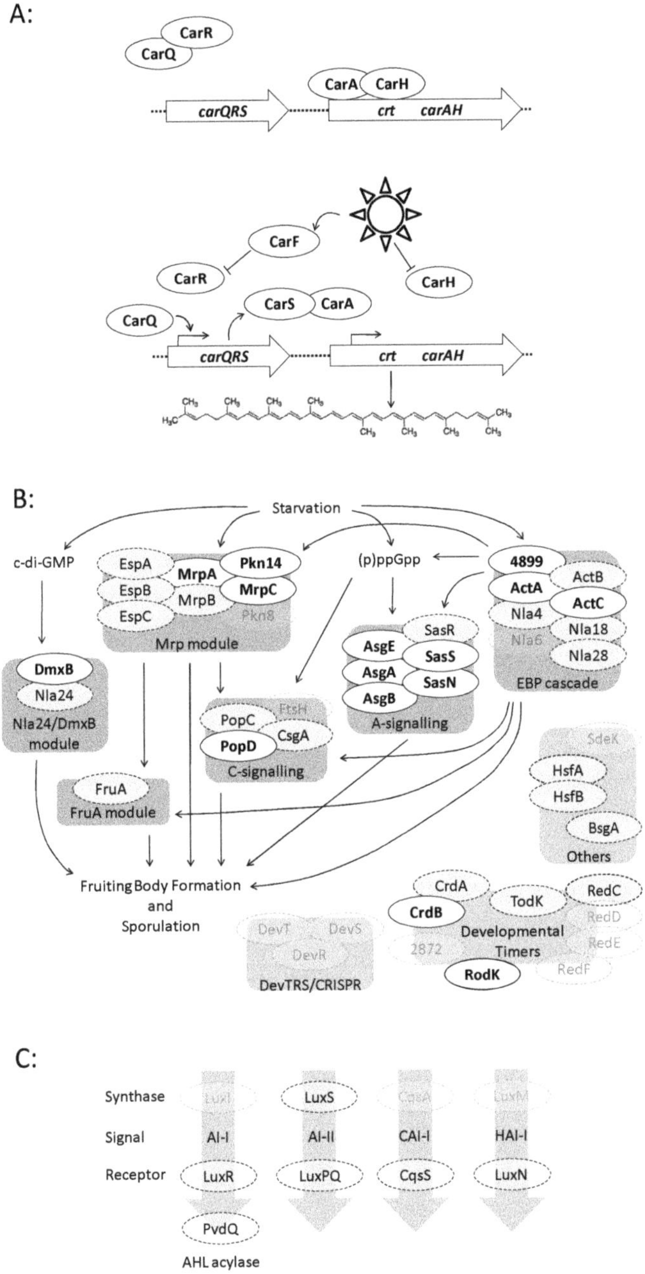

Figure 3. Conservation of signalling pathway proteins in Myxococcaceae. Regulatory proteins are shown as ovals. Positive regulation is shown with pointed arrows, and negative regulation with blunt-headed arrows. (**A**) Carotenoid production. In the dark (top), CarQ is held inactive by CarR, while CarA and CarH repress expression of the constitutively active *crt* promoter. In the light (bottom), CarH is directly inactivated while CarF inactivates CarR, releasing CarQ to direct transcription of

carQRS, producing CarS which binds to CarA, relieving repression of the crt genes, which encode enzymes for the biosynthesis of carotenoids (lycopene shown as an example). (**B**) Fruiting body formation. Gene products work together in modules (dark grey boxes). Starvation triggers the production of the secondary messengers c-di-GMP and (p)ppGpp and activates the EBP (enhancer binding protein) cascade and the Mrp module. The A-signalling, C-signalling, FruA and Nla24/DmxB modules are stimulated by secondary messengers and regulatory modules. Various 'development timer' proteins regulate the timing of fruiting, and the DevTRS/CRISPR module modulates the timing of sporulation. 'Other' proteins regulate fruiting, but their relationship to other modules is not clear. (**C**) Quorum signalling. Four common Gram-negative bacterial quorum signals (AI-I, AI-II, CAI-I and HAI-I), and their corresponding synthase and receptor/regulator proteins, are shown. Also shown is the quorum-quenching AHL acylase, PvdQ. Whether regulatory proteins are core, conserved, or dispensable, is indicated based on their pattern of evolutionary conservation. Core proteins are found at a constant number per genome across the Myxococcaceae and are highlighted in bold text. Conserved proteins are found in all groups of Myxococcaceal genomes, but in variable numbers, and are indicated with a pale grey background and dashed outline. Dispensable proteins are absent from some groups of Myxococcaeal genomes and are shown with a transparent background, and grey text.

3.6. Case Study 1: Carotenogenesis

Every protein of the carotenogenesis signalling pathway was found to exhibit the same pattern of conservation (Figure 3A), with a constant single orthologue in every group of genomes (Supplementary Table S2). Thus, every component of the pathway can be considered 'core', and essential for the functioning of the pathway across the Myxococcaceae. This is easy to rationalise since despite integrating proteins of several regulatory classes, the pathway is essentially linear, and losing any single component results in a defective response to toxic light.

3.7. Case Study 2: Fruiting Body Formation

In contrast to the carotenogenesis pathway, the regulation of fruiting body formation is dominated by TCS proteins, organised into a highly interconnected network of regulatory modules (Figure 3B). The main developmental regulators were categorised into the modules or processes described by Kroos [22], with an additional category of 'developmental timers' as defined by Diodati et al. [41], and then homologues were identified by BLAST.

Some modules were found to be composed entirely of core/conserved gene products, for example the FruA module (one protein) and the A-signalling module (six proteins), while several modules were largely core/conserved, but included the occasional dispensable protein (Supplementary Table S2). For instance, Pkn8 appears to be dispensable from the Mrp module (eight proteins) as previously noted by Kroos [22], the EBP (enhancer binding protein) module (eight proteins) can dispense with Nla6, while the C-signalling module (four proteins) is often found without an FtsH homologue. In the two-protein Nla24 module, Nla24 is dispensable and DmxB is core (the Nla24 module should therefore be renamed the DmxB module), while developmental timers are a mixture of core/conserved (five) and dispensable (four) proteins. The DevR, DevS and DevT CRISPR (clustered regularly interspaced short palindromic repeats)-related proteins which affect the timing of sporulation were all dispensable (as noted by Kroos [22]), consistent with the proposal that they do not regulate development per se, but instead increase phage-resistance during development.

In overview, it seems that all the modules involved in regulating development are found across the Myxococcaceae, suggesting that the general organisation of the developmental pathway is evolutionarily conserved. However, the modules frequently lack proteins that are required for proper development in *M. xanthus* DK1622, implying that the developmental network is evolutionarily robust—able to evolve to cope with both the loss of developmental genes and the integration of newly acquired/duplicated gene products.

3.8. Case Study 3: Quorum Signalling

In contrast to carotenogenesis and fruiting body formation, QS pathways are short, and operate independently of one another. Myxobacteria are generally thought not to engage in quorum signalling, as practised by other Gram-negative bacteria, which involves the secretion of an auto-inducer signalling molecule, which producing cells then respond to. Nevertheless, using query sequences from non-myxobacterial QS organisms, homologues of various QS proteins were detected in myxobacterial genomes by BLAST (Supplementary Table S2, Figure 3C).

No genomes encoded a HAI-I synthase homologue, but an AI-I synthase was found in *C. exiguus* AB016 and an AI-II synthase was found in *M. llanfairPGensis*. Surprisingly, more than three homologues of the CAI-I synthase CqsA were encoded in each genome. The sensors of most auto-inducers are HKs, so searches for homologues of the CqsS, LuxN and LuxQ sensors produced more than 100 hits in each genome. However, homologues of the LuxR TF sensor of AI-I were less abundant but were nonetheless conserved, with at least one homologue in each genome, except that of *C. llansteffanensis*. In addition, the PvdQ AHL (acyl homoserine lactone) acylase which quenches QS had conserved homologues in every genome. Thus, it seems that production of CAI-I is a common feature of these organisms, and occasional strains can produce additional QS molecules. The capacity to sense QS molecules is conserved, including in non-producing strains (eavesdroppers), as is the ability to quench the QS of other (potential prey) organisms.

4. Discussion

Myxobacterial genomes encode large numbers of signalling proteins; however, within a genus, they also have very small core genomes due to the large proportion of accessory genes in each genome [7–9]. Previous analysis of conservation of myxobacterial TCS genes suggested that gene gain/loss was one of the most frequent types of mutational events experienced by TCS genes [3]. Nevertheless, we would expect that some TCS genes belong to the core genome and are indispensable, while other TCS genes would belong to the accessory pan-genome and would be absent from some organisms. We therefore investigated the conservation of regulatory gene family members within groups of myxobacterial genomes, and also assessed conservation of regulators associated with key myxobacterial behaviours.

The numbers of regulatory proteins of different families/classes is remarkably constant between genomes within a group of related organisms, suggesting that they are disproportionately represented in the core genome compared to 'typical' genes (Figure 2). TCS genes seem to be even more enriched in the core genome compared to OCSs and TFs, which perhaps reflects the large numbers of TCSs in myxobacterial genomes. Because of their shared domain architectures and mechanisms of phosphotransfer, multiple TCS signalling pathways can be integrated into sophisticated regulatory modules and networks [42]. Potentially, this might reduce the loss of individual TCS genes from genomes, with selection instead acting at the level of the whole network or module.

Fruiting body formation in *M. xanthus* is regulated by a modular network dominated by TCS proteins. However, selection does not seem to be at the level of the module. The only module that is either present or absent from different genomes in its entirety is the DevTRS/CRISPR module (Supplementary Table S2), which is thought to primarily resist phage infection during sporulation, with only secondary effects on the timing of sporulation [22]. The other regulatory modules are always present in a genome, but in every case, some individual components are conserved, while others are dispensable (Figure 3).

Robustness is a global property of modular biological networks, and the lack of conservation of 'key' fruiting regulatory proteins in species/strains which are proficient in fruiting implies the myxobacterial developmental network is evolutionarily robust. The impact of mutational loss can be reduced by the architecture of signalling networks [43]. It has long been recognised that robustness is an emergent property of certain network architectures. In particular, modularity is an organising principle allowing the evolution of both robustness and computational complexity in gene regulatory networks [44,45].

The ease with which suppressor and bypass mutations of developmental gene mutants can be isolated supports the notion that the developmental network is evolutionarily robust. Examples abound, but one good example involves the non-coding RNA Pxr, which inhibits the initiation of fruiting body development in the presence of nutrients [46]. In one study, a mutant strain unable to relieve Pxr inhibition in response to starvation could be restored to developmental proficiency by mutations within three separate genes, *pxr* and two positive regulators of *pxr* expression, leading the authors to conclude that reversion of developmental defects could be commonplace [47]. In another example, a third separate bypass suppressor mutation of the protease gene *bsgA* was mapped to an operon encoding RNase D and an aminopeptidase [48].

The network must also be able to incorporate newly acquired or duplicated genes. Potentially, subtle changes in phenotype due to acquisition of a new gene might confer enough of a selective advantage to promote retention of that gene. Gene duplication seems to have contributed to the large expansion in the size of myxobacterial genomes compared to the other Deltaproteobacterial Orders, with EBPs and TCSs notably prevalent [36], while acquisition by horizontal transfer might explain the origin of more than 20% of contemporary myxobacterial genes [49]. It is possible that TCSs are particularly abundant in the fruiting regulatory network because they are better able than other regulators to tolerate changes to network architecture and to engage in complex interactions with multiple partner regulators.

In contrast to the fruiting body formation network, the Car system of *M. xanthus* is essentially a linear signalling pathway, reliant on the sequential action of different categories of regulators. Unsurprisingly, all Car pathway genes are conserved in all genomes analysed (Figure 3, Supplementary Table S2). Such a pattern of conservation implies that the Car pathway regulates a phenotype with a strong selective advantage. Absence of carotenoid biosynthesis would make cells sensitive to singlet oxygen-mediated damage, resulting in death and a clear selective pressure. But, presumably, the metabolic costs of producing photoprotective carotenoids constitutively are also high enough to make retention of the signalling pathway evolutionarily favourable.

Myxobacteria are generally considered to not produce the AHLs that mediate QS in diverse Gram-negative bacteria, although recently, a cryptic myxobacterial gene resembling an AHL synthase (*agpI*) was identified in the myxobacterium *Archangium gephyra* [50]. The *agpI* gene was found to be able to induce production of AHLs in *Escherichia coli*, suggesting that AgpI may play a role in disrupting communication between prey. In addition, exogenously added AHLs have been found to promote the predatory behaviours of *M. xanthus* [26], suggesting that myxobacteria might eavesdrop on their prey. The conservation of an AHL acylase suggests that active disruption of prey AHL-mediated QS might be a common behaviour of predatory myxobacteria. Conservation of CAI-I synthase homologues suggests that myxobacteria may communicate amongst themselves via this form of QS, while the occasional strain may also be able to use alternative QS molecules (Figure 3, Supplementary Table S2). CAI-I signalling has been most commonly associated with marine bacteria, and diverse chemical variants of CAI-I have been described [25,51]. We predict that myxobacteria generally produce CAI-I variants and note that 90% of *Corallococcus* spp. type strains are predicted by antiSMASH 5 to produce homoserine lactones and/or butyrolactones, with the latter being QS molecules associated with the phylum Actinobacteria [7,52]. Further studies on QS in myxobacteria are needed to unravel what is likely to be a pervasive but idiosyncratic feature of their biology.

Clearly, different types of signalling pathways and behaviours exhibit differing patterns of gene conservation. For some pathways (e.g., the Car pathway), every component gene is highly conserved, some (e.g., fruiting body formation) are largely conserved but particular genes are dispensable, while others are present sporadically within a taxon (e.g., AI-I and AI-II synthases). As well as the structure of the pathway (modular vs. linear) and its evolutionary robustness, the pattern of conservation is also likely to be affected by the number of genetic loci over which the regulatory genes are found.

For example, pathways present at single loci (e.g., QS pathways) can be acquired/lost in their entirety by single mutagenic events, whereas networks comprising large numbers of components encoded at multiple loci (e.g., fruiting body formation) are more likely to gain/lose sub-components rather than entire modules.

The availability of genome sequences means that knowledge gained by researching the molecular genetics of one model bacterium can be easily translated onto another organism by comparing their gene sets. It will be particularly interesting to extend these analyses to myxobacteria beyond the Myxococcaceae when more genomes become available. However, there are important caveats that must be appreciated when doing so, or we risk over-interpreting the significance of homologue presence/absence/variation [53], especially if using draft rather than complete genome sequences. Specifically, it seems that in myxobacteria, even if a regulatory pathway confers a selective advantage, individual genes involved in that process will likely only be evolutionarily conserved if the pathway is linear with a small-to-medium number of genes. For complex regulatory processes involving large numbers of genes (e.g., fruiting body formation), just because a gene is essential for that process in a model organism like *M. xanthus* DK1622, it cannot be assumed that it will also be required, or fulfilling the same role, in other members of that species/genus.

Supplementary Materials: The following are available online at http://www.mdpi.com/2076-2607/8/11/1739/s1, Table S1: Regulatory proteins encoded in 41 myxobacterial genomes, plus that of the *M. xanthus* type strain DSM 16526. Table S2: Pattern of conservation and the number of homologues identified in each Myxococcaceal genome when queried with regulatory proteins involved in carotenogenesis, fruiting body formation and quorum signalling.

Author Contributions: Conceptualization, D.E.W; methodology, D.E.W. and A.Z.; investigation, D.E.W. and A.Z.; writing—review and editing, D.E.W. and A.Z. All authors have read and agreed to the published version of the manuscript.

Funding: This research received no external funding.

Conflicts of Interest: The authors declare no conflict of interest.

References

1. Pérez, J.; Castañeda-García, A.; Jenke-Kodama, H.; Müller, R.; Muñoz-Dorado, J. Eukaryotic-like protein kinases in the prokaryotes and the myxobacterial kinome. *Proc. Natl. Acad. Sci. USA* **2008**, *105*, 15950–15955. [CrossRef]
2. Whitworth, D.E.; Cock, P.J.A. Two-component systems of the myxobacteria: Structure, diversity and evolutionary relationships. *Microbiology* **2008**, *154 Pt 2*, 360–372. [CrossRef]
3. Whitworth, D.E. Genome-wide analysis of myxobacterial two-component systems: Genome relatedness and evolutionary changes. *BMC Genom.* **2015**, *16*, 780. [CrossRef]
4. Garcia, R.; Gerth, K.; Stadler, M.; Dogma, I.J., Jr.; Müller, R. Expanded phylogeny of myxobacteria and evidence for cultivation of the 'unculturables'. *Mol. Phylogenet. Evol.* **2010**, *57*, 878–887. [CrossRef]
5. Sharma, G.; Narwani, T.; Subramanian, S. Complete Genome Sequence and Comparative Genomics of a Novel Myxobacterium Myxococcus hansupus. *PLoS ONE* **2016**, *11*, e0148593. [CrossRef] [PubMed]
6. Livingstone, P.G.; Ingleby, O.; Girdwood, S.; Cookson, A.R.; Morphew, R.M.; Whitworth, D.E. Predatory Organisms with Untapped Biosynthetic Potential: Descriptions of Novel *Corallococcus* Species *C. aberystwythensis* sp. nov., *C. carmarthensis* sp. nov., *C. exercitus* sp. nov., *C. interemptor* sp. nov., *C. llansteffanensis* sp. nov., *C. praedator* sp. nov., *C. sicarius* sp. nov., and *C. terminator* sp. nov. *Appl. Environ. Microbiol.* **2020**, *86*, e01931-19. [PubMed]
7. Chambers, J.; Sparks, N.; Sydney, N.; Livingstone, P.G.; Cookson, A.R.; Whitworth, D.E. Comparative genomics and pan-genomics of the Myxococcaceae, including a description of five novel species: *Myxococcus eversor* sp. nov., *Myxococcus llanfairpwllgwyngyllgogerychwyrndrobwllllantysiliogogogochensis* sp. nov., *Myxococcus vastator* sp. nov., *Pyxidicoccus caerfyrddinensis* sp. nov., and *Pyxidicoccus trucidator* sp. nov. *Genome Biol. Evol.* **2020**, in press.

8. Livingstone, P.G.; Morphew, R.M.; Whitworth, D.E. Genome Sequencing and Pan-Genome Analysis of 23 *Corallococcus* spp. Strains Reveal Unexpected Diversity, with Particular Plasticity of Predatory Gene Sets. *Front. Microbiol.* **2018**, *9*, 3187. [CrossRef]
9. Zwarycz, A.S.; Livingstone, P.G.; Whitworth, D.E. Within-species variation in OMV cargo proteins: The *Myxococcus xanthus* OMV pan-proteome. *Mol. Omics* **2020**. [CrossRef]
10. Giovannoni, S.J.; Cameron Thrash, J.; Temperton, B. Implications of streamlining theory for microbial ecology. *ISME J.* **2014**, *8*, 1553–1565. [CrossRef]
11. Sutton, D.; Livingstone, P.G.; Furness, E.; Swain, M.T.; Whitworth, D.E. Genome-Wide Identification of Myxobacterial Predation Genes and Demonstration of Formaldehyde Secretion as a Potentially Predation-Resistant Trait of *Pseudomonas aeruginosa*. *Front. Microbiol.* **2019**, *10*, 2650. [CrossRef]
12. Galbis-Martínez, M.; Padmanabhan, S.; Murillo, F.J.; Elías-Arnanz, M. CarF mediates signaling by singlet oxygen, generated via photoexcited protoporphyrin IX in *Myxococcus xanthus* light-induced carotenogenesis. *J. Bacteriol.* **2012**, *194*, 1427–1436. [CrossRef]
13. Browning, D.F.; Whitworth, D.E.; Hodgson, D.A. Light-induced carotenogenesis in *Myxococcus xanthus*: Functional characterization of the ECF sigma factor CarQ and antisigma factor CarR. *Mol. Microbiol.* **2003**, *48*, 237–251. [CrossRef]
14. Whitworth, D.E.; Bryan, S.J.; Berry, A.E.; McGowan, S.J.; Hodgson, D.A. Genetic dissection of the light-inducible carQRS promoter region of *Myxococcus xanthus*. *J. Bacteriol.* **2004**, *186*, 7836–7846. [CrossRef]
15. Whitworth, D.E.; Hodgson, D.A. Light-induced carotenogenesis in *Myxococcus xanthus*: Evidence that CarS acts as an anti-repressor of CarA. *Mol. Microbiol.* **2001**, *42*, 809–819. [CrossRef]
16. López-Rubio, J.J.; Elías-Arnanz, M.; Padmanabhan, S.; Murillo, F.J. A repressor-antirepressor pair links two loci controlling light-induced carotenogenesis in *Myxococcus xanthus*. *J. Biol. Chem.* **2002**, *277*, 7262–7270. [CrossRef]
17. León, E.; Navarro-Avilés, G.; Santiveri, C.M.; Flores-Flores, C.; Rico, M.; González, C.; Murillo, F.J.; Elías-Arnanz, M.; Jiménez, M.A.; Padmanabhan, S. A bacterial antirepressor with SH3 domain topology mimics operator DNA in sequestering the repressor DNA recognition helix. *Nucleic Acids Res.* **2010**, *38*, 5226–5241. [CrossRef]
18. Padmanabhan, S.; Jost, M.; Drennan, C.L.; Elías-Arnanz, M. A New Facet of Vitamin B(12): Gene Regulation by Cobalamin-Based Photoreceptors. *Annu. Rev. Biochem.* **2017**, *86*, 485–514. [CrossRef]
19. Sarwar, Z.; Garza, A.G. Two-Component Signal Transduction Systems That Regulate the Temporal and Spatial Expression of *Myxococcus xanthus* Sporulation Genes. *J. Bacteriol.* **2015**, *198*, 377–385. [CrossRef]
20. Bretl, D.J.; Kirby, J.R. Molecular Mechanisms of Signaling in *Myxococcus xanthus* Development. *J. Mol. Biol.* **2016**, *428*, 3805–3830. [CrossRef]
21. Muñoz-Dorado, J.; Moraleda-Muñoz, A.; Marcos-Torres, F.J.; Contreras-Moreno, F.J.; Martin-Cuadrado, A.B.; Schrader, J.M.; Higgs, P.I.; Pérez, J. Transcriptome dynamics of the *Myxococcus xanthus* multicellular developmental program. *Elife* **2019**, *8*, e50374. [CrossRef]
22. Kroos, L. Highly Signal-Responsive Gene Regulatory Network Governing *Myxococcus* Development. *Trends Genet.* **2017**, *33*, 3–15. [CrossRef]
23. Konovalova, A.; Wegener-Feldbrügge, S.; Søgaard-Andersen, L. Two intercellular signals required for fruiting body formation in *Myxococcus xanthus* act sequentially but non-hierarchically. *Mol. Microbiol.* **2012**, *86*, 65–81. [CrossRef]
24. Ng, W.L.; Bassler, B.L. Bacterial quorum-sensing network architectures. *Annu. Rev. Genet.* **2009**, *43*, 197–222. [CrossRef] [PubMed]
25. Papenfort, K.; Bassler, B.L. Quorum sensing signal-response systems in Gram-negative bacteria. *Nat. Rev. Microbiol.* **2016**, *14*, 576–588. [CrossRef]
26. Lloyd, D.G.; Whitworth, D.E. The Myxobacterium *Myxococcus xanthus* Can Sense and Respond to the Quorum Signals Secreted by Potential Prey Organisms. *Front. Microbiol.* **2017**, *8*, 439. [CrossRef]
27. Whitworth, D.E.; Swain, M.T. A survey of non-coding RNAs in the social and predatory myxobacterium *Myxococcus xanthus* DK1622. *Mol. Omics* **2020**. [CrossRef]
28. Livingstone, P.G.; Morphew, R.M.; Whitworth, D.E. Myxobacteria Are Able to Prey Broadly upon Clinically-Relevant Pathogens, Exhibiting a Prey Range Which Cannot Be Explained by Phylogeny. *Front. Microbiol.* **2017**, *8*, 1593. [CrossRef]

29. Li, H.; Durbin, R. Fast and accurate short read alignment with Burrows-Wheeler transform. *Bioinformatics* **2009**, *25*, 1754–1760. [CrossRef]
30. Bankevich, A.; Nurk, S.; Antipov, D.; Gurevich, A.A.; Dvorkin, M.; Kulikov, A.S.; Lesin, V.M.; Nikolenko, S.I.; Pham, S.; Prjibelski, A.D.; et al. SPAdes: A new genome assembly algorithm and its applications to single-cell sequencing. *J. Comput. Biol.* **2012**, *19*, 455–477. [CrossRef]
31. Wood, D.E.; Salzberg, S.L. Kraken: Ultrafast metagenomic sequence classification using exact alignments. *Genome Biol.* **2014**, *15*, R46. [CrossRef]
32. Haft, D.H.; DiCuccio, M.; Badretdin, A.; Brover, V.; Chetvernin, V.; O'Neill, K.; Li, W.; Chitsaz, F.; Derbyshire, M.K.; Gonzales, N.R.; et al. RefSeq: An update on prokaryotic genome annotation and curation. *Nucleic Acids Res.* **2018**, *46*, D851–D860. [CrossRef]
33. Barakat, M.; Ortet, P.; Whitworth, D.E. P2RP: A Web-based framework for the identification and analysis of regulatory proteins in prokaryotic genomes. *BMC Genom.* **2013**, *14*, 269. [CrossRef] [PubMed]
34. Ortet, P.; De Luca, G.; Whitworth, D.E.; Barakat, M. P2TF: A comprehensive resource for analysis of prokaryotic transcription factors. *BMC Genom.* **2012**, *13*, 628. [CrossRef] [PubMed]
35. Ortet, P.; Whitworth, D.E.; Santaella, C.; Achouak, W.; Barakat, M. P2CS: Updates of the prokaryotic two-component systems database. *Nucleic Acids Res.* **2015**, *43*, D536–D541. [CrossRef] [PubMed]
36. Goldman, B.S.; Nierman, W.C.; Kaiser, D.; Slater, S.C.; Durkin, A.S.; Eisen, J.A.; Ronning, C.M.; Barbazuk, W.B.; Blanchard, M.; Field, C.; et al. Evolution of sensory complexity recorded in a myxobacterial genome. *Proc. Natl. Acad. Sci. USA* **2006**, *103*, 15200–15205. [CrossRef] [PubMed]
37. Treuner-Lange, A.; Bruckskotten, M.; Rupp, O.; Goesmann, A.; Søgaard-Andersen, L. Complete Genome Sequence of the Fruiting Myxobacterium *Myxococcus macrosporus* Strain DSM 14697, Generated by PacBio Sequencing. *Genome Announc.* **2017**, *5*, e01127-17. [CrossRef]
38. Huntley, S.; Kneip, S.; Treuner-Lange, A.; Søgaard-Andersen, L. Complete genome sequence of *Myxococcus stipitatus* strain DSM 14675, a fruiting myxobacterium. *Genome Announc.* **2013**, *1*, e0010013. [CrossRef]
39. Huntley, S.; Zhang, Y.; Treuner-Lange, A.; Kneip, S.; Sensen, C.W.; Søgaard-Andersen, L. Complete genome sequence of the fruiting myxobacterium *Corallococcus coralloides* DSM 2259. *J. Bacteriol.* **2012**, *194*, 3012–3013. [CrossRef]
40. El-Gebali, S.; Mistry, J.; Bateman, A.; Eddy, S.R.; Luciani, A.; Potter, S.C.; Qureshi, M.; Richardson, L.J.; Salazar, G.A.; Smart, A.; et al. The Pfam protein families database in 2019. *Nucleic Acids Res.* **2019**, *47*, D427–D432. [CrossRef]
41. Diodati, M.E.; Gill, R.E.; Plamann, L.; Singer, M. Initiation and early developmental events. In *Myxobacteria: Multicellularity and Differentiation*; Whitworth, D.E., Ed.; American Society for Microbiology: Washington, DC, USA, 2008; pp. 43–76.
42. Francis, V.I.; Porter, S.L. Multikinase Networks: Two-Component Signaling Networks Integrating Multiple Stimuli. *Annu. Rev. Microbiol.* **2019**, *73*, 199–223. [CrossRef]
43. Payne, J.L.; Wagner, A. Mechanisms of mutational robustness in transcriptional regulation. *Front. Genet.* **2014**, *6*, 322. [CrossRef]
44. Silva-Rocha, R.; de Lorenzo, V. Noise and robustness in prokaryotic regulatory networks. *Annu. Rev. Microbiol.* **2010**, *64*, 257–275. [CrossRef]
45. Agarwal, S. Systems approaches in understanding evolution and evolvability. *Prog. Biophys. Mol. Biol.* **2013**, *113*, 369–374. [CrossRef]
46. Yu, Y.T.; Yuan, X.; Velicer, G.J. Adaptive evolution of ansRNA that controls *Myxococcus* development. *Science* **2010**, *328*, 993. [CrossRef]
47. Yu, Y.N.; Kleiner, M.; Velicer, G.J. Spontaneous Reversions of an Evolutionary Trait Loss Reveal Regulators of a Small RNA That Controls Multicellular Development in Myxobacteria. *J. Bacteriol.* **2016**, *198*, 3142–3151. [CrossRef] [PubMed]
48. Cusick, J.K.; Hager, E.; Gill, R.E. Identification of a mutant locus that bypasses the BsgA protease requirement for social development in *Myxococcus xanthus*. *FEMS Microbiol. Lett.* **2015**, *362*, 1–8. [CrossRef]
49. Goldman, B.; Bhat, S.; Shimkets, L.J. Genome evolution and the emergence of fruiting body development in *Myxococcus xanthus*. *PLoS ONE* **2007**, *2*, e1329. [CrossRef]
50. Albataineh, H.; Duke, M.; Misra, S.K.; Sharp, J.S.; Cole Stevens, D. Cryptic, solo acyl homoserine lactone synthase from predatory myxobacterium suggests beneficial contribution to prey quorum signaling. *bioRxiv* **2020**. [CrossRef]

51. Henke, J.M.; Bassler, B.L. Three parallel quorum-sensing systems regulate gene expression in *Vibrio harveyi*. *J. Bacteriol.* **2004**, *186*, 6902–6914. [CrossRef]
52. Polkade, A.V.; Mantri, S.S.; Patwekar, U.J.; Jangid, K. Quorum Sensing: An Under-Explored Phenomenon in the Phylum Actinobacteria. *Front. Microbiol.* **2016**, *7*, 131. [CrossRef]
53. Whitworth, D.E. Genomes and knowledge—A questionable relationship? *Trends Microbiol.* **2008**, *16*, 512–519. [CrossRef]

Publisher's Note: MDPI stays neutral with regard to jurisdictional claims in published maps and institutional affiliations.

microorganisms

MDPI

Review

Light-Triggered Carotenogenesis in *Myxococcus xanthus*: New Paradigms in Photosensory Signaling, Transduction and Gene Regulation

S. Padmanabhan [1,*], Antonio J. Monera-Girona [2], Ricardo Pérez-Castaño [2], Eva Bastida-Martínez [2], Elena Pajares-Martínez [2], Diego Bernal-Bernal [1], María Luisa Galbis-Martínez [2], María Carmen Polanco [2], Antonio A. Iniesta [2], Marta Fontes [2] and Montserrat Elías-Arnanz [2,*]

1 Instituto de Química Física Rocasolano, Consejo Superior de Investigaciones Científicas, 28006 Madrid, Spain; diego.bernal@iqfr.csic.es
2 Departamento de Genética y Microbiología, Área de Genética (Unidad Asociada al IQFR-CSIC), Facultad de Biología, Universidad de Murcia, 30100 Murcia, Spain; aj.moneragirona@um.es (A.J.M.-G.); ricardo.perez@um.es (R.P.-C.); eva.bastida@um.es (E.B.-M.); elena.pajares@um.es (E.P.-M.); mgalbis@um.es (M.L.G.-M.); mpolanco@um.es (M.C.P.); ainiesta@um.es (A.A.I.); mfontes@um.es (M.F.)
* Correspondence: padhu@iqfr.csic.es (S.P.); melias@um.es (M.E.-A.)

Abstract: Myxobacteria are Gram-negative δ-proteobacteria found predominantly in terrestrial habitats and often brightly colored due to the biosynthesis of carotenoids. Carotenoids are lipophilic isoprenoid pigments that protect cells from damage and death by quenching highly reactive and toxic oxidative species, like singlet oxygen, generated upon growth under light. The model myxobacterium *Myxococcus xanthus* turns from yellow in the dark to red upon exposure to light because of the photoinduction of carotenoid biosynthesis. How light is sensed and transduced to bring about regulated carotenogenesis in order to combat photooxidative stress has been extensively investigated in *M. xanthus* using genetic, biochemical and high-resolution structural methods. These studies have unearthed new paradigms in bacterial light sensing, signal transduction and gene regulation, and have led to the discovery of prototypical members of widely distributed protein families with novel functions. Major advances have been made over the last decade in elucidating the molecular mechanisms underlying the light-dependent signaling and regulation of the transcriptional response leading to carotenogenesis in *M. xanthus*. This review aims to provide an up-to-date overview of these findings and their significance.

Keywords: photoreceptor; photosensitizer; photoregulation; singlet oxygen; plasmalogens; CarF; vitamin B_{12}; CarH; ECF-sigma; CarD-CdnL

Citation: Padmanabhan, S.; Monera-Girona, A.J.; Pérez-Castaño, R.; Bastida-Martínez, E.; Pajares-Martínez, E.; Bernal-Bernal, D.; Galbis-Martínez, M.L.; Polanco, M.C.; Iniesta, A.A.; Fontes, M.; et al. Light-Triggered Carotenogenesis in *Myxococcus xanthus*: New Paradigms in Photosensory Signaling, Transduction and Gene Regulation. *Microorganisms* **2021**, *9*, 1067. https://doi.org/10.3390/microorganisms9051067

Academic Editor: David Whitworth

Received: 29 April 2021
Accepted: 12 May 2021
Published: 15 May 2021

1. Introduction

Light is an important and ubiquitous signal in terrestrial and aquatic ecosystems, and the ability to sense, respond and adapt to light is crucial for most living organisms, including bacteria. Photosynthetic bacteria capture and convert light, an essential energy source, to chemical energy for cellular utilization, but light is also important for several other cellular processes in both phototrophic and non-phototrophic bacteria [1–5]. Thus, light is linked to many bacterial responses such as phototaxis, development, virulence, circadian rhythms and UV-induced DNA damage repair [4–7]. However, light can be harmful and cause cell damage and death. This stems from excitation of photosensitizing biomolecules, such as porphyrins, chlorophyll or flavins, to generate highly reactive oxygen species (ROS) like singlet oxygen ($^1O_2$), superoxides, peroxides and hydroxyl radicals that can destroy cellular DNA, protein and lipid components [4,8–11]. Consequently, bacteria have evolved ingenious mechanisms and machineries to mount a protective response to counter photooxidative stress.

A commonly used defense mechanism against photooxidative damage is through the biosynthesis of carotenoids, which quench and dissipate as heat the excess energy of $^1O_2$ and other ROS produced upon illumination [4,8,9,11–13]. Carotenoids constitute a major class of lipophilic isoprenoid derivatives that are characterized by an extended, typically all-*trans*, conjugated polyene chain (usually C_{40} and some C_{50}, C_{45} and C_{30} terpenes) with acyclic, monocyclic or bicyclic ends. Their oxygenated (hydroxy, aldehyde, keto, carboxyl, methoxy, epoxy, oxy and glycosidic) derivatives are called xanthophylls. Most carotenoids are richly colored (light yellow to deep red), since they absorb blue-violet light (400–500 nm range) owing to their extended conjugated double bonds that also determine the molecular conformation and reactivity [13]. Carotenoids also fulfill biological roles other than in photoprotection, such as in photosynthetic light harvesting, signaling and as precursors of photosensory molecules and hormones [13,14].

Carotenoid biosynthesis de novo occurs in all photosynthetic organisms (plants, algae or bacteria) and in many non-photosynthetic fungi, archaea and bacteria, whereas animals, save some strikingly few exceptions, do not synthesize carotenoids but obtain them exogenously [13,14]. Given that carotenoids are in the frontline of the defense against photooxidative stress, light and oxygen-related species like $^1O_2$ are among the principal environmental factors involved in signaling and triggering carotenoid biosynthesis. This has been amply demonstrated in several studies from plants [15,16] and fungi [17,18] to bacteria [4,8,9,12]. Light-induced carotenogenesis and its regulation in the Gram-negative soil bacterium *M. xanthus* is undoubtedly one of the best studied and characterized among bacteria. We last reviewed this topic over a decade ago when many questions remained open [4,12]. Since then, considerable progress has been achieved largely from work in our group on the mechanistic, structural and photochemical aspects of light-regulated carotenogenesis in *M. xanthus*. Our work has uncovered new and large protein families, such as an entirely new class of photoreceptors with their novel mode of action that we specifically reviewed elsewhere [19–22]. It has also revealed the participation of "eukaryotic-like" proteins, including one found in *M. xanthus* and related myxobacteria, but absent in the vast majority of other bacteria, that turned out to be a long-sought human enzyme conserved across metazoa [23]. Our present review aims to provide a timely update of these findings, from signal reception and transduction to the transcriptional regulation underlying the photooxidative stress response and carotenoid biosynthesis in *M. xanthus*, and to discuss their mechanistic and evolutionary significance.

2. Biosynthesis of Carotenoids

Carotenoid biosynthesis occurs via a well-established and largely conserved pathway involving a number of genes and their products [13,14]. The pathway is considered to begin with the condensation of the universal five-carbon (C_5) isoprenoid precursors isopentenyl diphosphate (IPP) and dimethylallyl diphosphate (DMAPP), themselves products of either the mevalonate (MVA) pathway (see Figure 1) or the non-mevalonate 2C-methyl-D-erythritol-4-phosphate (MEP) pathway [13,14,24]. Most bacteria and plastids are equipped with the MEP pathway, the MVA pathway is prevalent in animals, archaea, fungi and some bacteria including *M. xanthus* and the majority of myxobacteria, while plants and some select bacterial species use both pathways [13,14,24]. Condensation of IPP and DMAPP, the first committed and usually rate-controlling step in the core carotenoid biosynthesis pathway, produces geranylgeranyl diphosphate, two molecules of which then condense to generate the colorless C_{40} isoprenoid phytoene. A series of phytoene isomerization and desaturation steps generates the red carotenoid lycopene, from which carotenes and xanthophylls are produced in further desaturation, isomerization and hydroxylation reactions. Carotenoid biosynthesis and its regulation at levels from transcription, which is among the earliest and most crucial steps, to post-translation, degradation and feedback have been studied in many organisms [13,14]. Here, we discuss our current understanding of the *M. xanthus* carotenoid biosynthesis pathway, the structural and regulatory genes involved and how their transcription is induced and regulated.

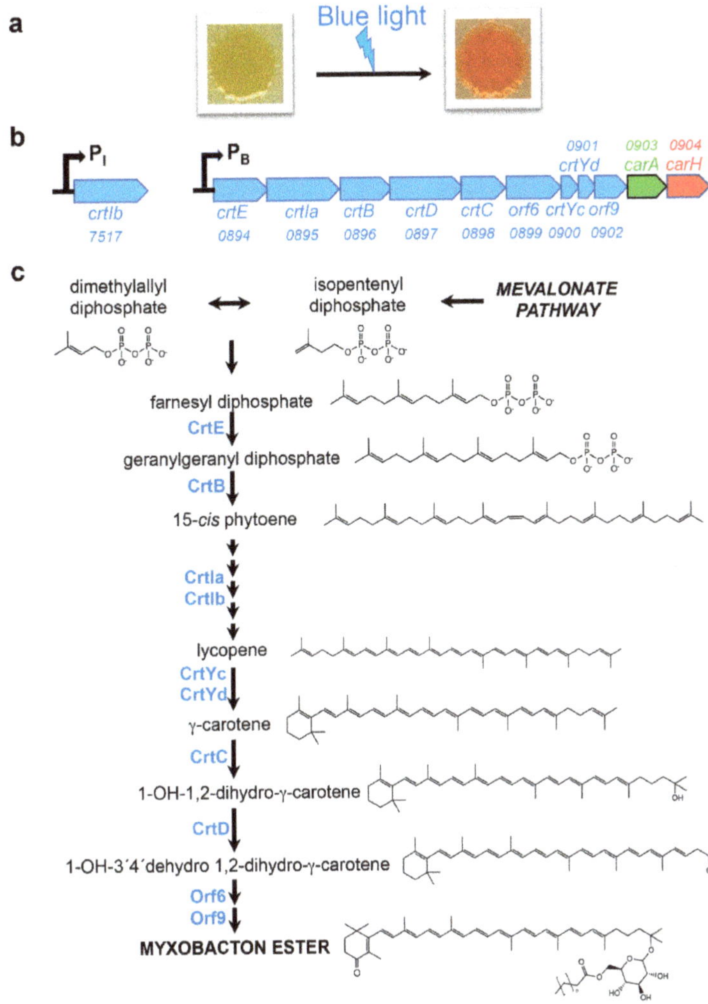

Figure 1. Carotenoid biosynthesis pathway and genes in *M. xanthus*. (**a**) *M. xanthus* colony color in the dark and in the light. Wild-type strains are yellow in the dark and red when exposed to blue light. (**b**) Structural genes for carotenogenesis characterized in *M. xanthus*. The *carB* locus encodes nine structural genes for carotenoid synthesis and two transcription regulatory factors, CarA and CarH, expressed from the primary σ^A-dependent P_B promoter. The isolated structural *crtIb* gene is expressed from a promoter that depends on the ECF-σ factor CarQ (see text). The four-digit number above or below the corresponding gene indicates the original genome locus tag (MXAN_xxxx) of each of these genes. (**c**) Carotenoid synthesis pathway derived from the mevalonate (MVA) pathway in *M. xanthus*, with enzymes and products indicated.

3. A Brief History of Early Findings in *M. xanthus* Light-Induced Carotenogenesis

M. xanthus cells are yellow (Figure 1a) in the dark due to noncarotenoid, light-sensitive pigments that were identified and named DKxanthenes just fifteen years ago [25]. Four decades earlier, Burchard, Dworkin and coworkers reported that *M. xanthus* cells, when grown in the light, suffered photolysis or developed an orange/red color attributed to carotenoids and resisted photolysis, with the extent of illumination and growth phase determining the accumulation of carotenoids [26]. The action spectrum for photoinduction of carotenoids mirrored those for photolysis and for the absorption spectrum of protoporphyrin IX (PPIX), a hydrophobic cyclic tetrapyrrole and immediate precursor of heme in its biosynthesis, which accumulates in the *M. xanthus* cell membrane especially during stationary phase [26,27]. Photoinduction of carotenogenesis was maximal under blue light (405–410 nm), with lower maxima in the green light region (510–580 nm). Blue light excites the photosensitizer PPIX to ^{3}PPIX, a very reactive high-energy triplet state that can directly cause cell damage or transfer its energy to other molecules [8,9]. Energy transfer from ^{3}PPIX to molecular oxygen generates $^1O_2$, an extremely reactive ROS that is relatively long-lived and diffusible in membrane environments [28]. Light-generated $^1O_2$ was therefore proposed as the signal for carotenoid biosynthesis in *M. xanthus* [29], and later validated experimentally [30].

Two decades after these early findings, isolation and genetic analysis of *M. xanthus* spontaneous mutants, or ones generated by chemical, UV or Tn*5-lac* insertions, helped identify key genetic loci involved in light-induced carotenogenesis and established it as a transcriptional response [31–34]. The distinctive color change from yellow in the dark to red in the light (Figure 1a) due to light-induced carotenogenesis (the wild-type Car$^+$ phenotype) provided a valuable visual tool for facile genetic analysis. This helped identify mutants that synthesize carotenoids constitutively (Car^C) and are always orange/red, and these mutations mapped to two loci, *carA* and *carR*, which were inferred to encode negative regulators. On the other hand, mutants that never turn red in the light (Car$^-$) were also identified and these mutations mapped to two loci encoding carotenogenic enzymes, or to various loci encoding putative positive regulators [32–37]. In these and subsequent studies, the loci were further mapped, epistatic relationships between them established, and the stage set for cloning, sequencing, gene expression assays and chemical analysis of carotenoids. This uncovered most of the structural and regulatory genes, and more recent biochemical, biophysical, genome-level and high-resolution structural analyses have provided profound insights into the molecular mechanisms underlying this light response.

4. Structural Genes Encoding *M. xanthus* Light-Induced Carotenoid Biosynthetic Enzymes

Genetic analysis, cloning and sequencing of the loci involved revealed that structural genes encoding the carotenoid synthesis enzymes (gene names usually prefixed *crt*) were located at the unlinked *carB* and *carC* loci [31–34,38–40]. The *carB* locus groups nine structural genes and two regulatory genes organized as *crtE-crtIa-crtB-crtD-crtC-orf6-crtYc-crtYd-orf9-carA-carH*, and the *carC* locus corresponds to a single gene, *crtIb* (Figure 1b). These annotations were based on analysis of sequence and of carotenoids accumulated in different mutants, as well as on heterologous expression in *E. coli* [33,38–42]. Genes *orf6* and *orf9* at the *carB* locus may also be structural ones whose functions remain to be established experimentally, while *carA* and *carH* encode transcription factors that regulate expression of *carB* genes (see below). The proposed carotenoid biosynthesis pathway for *M. xanthus* and the enzyme(s) involved in each step (Figure 1c) leads to synthesis of the final product, myxobacton ester, a monocyclic carotenoid with a keto group in the ring at one end of the molecule and a glycosyl group esterified to a straight-chain fatty acid at the other [43].

CrtE (geranylgeranyl pyrophosphate synthase), encoded by the first gene at the *carB* locus, catalyzes conversion of farnesyl diphosphate to geranylgeranyl diphosphate, two molecules of which condense to phytoene through the action of CrtB (phytoene synthase). The colorless phytoene is isomerized and transformed in four successive dehydrogenation steps to the red acyclic lycopene by the synergistic action of the CrtIa and CrtIb phytoene dehydrogenases [38,39,41]. Lycopene is cyclized at one end to monocyclic γ-carotene by the concerted action of CrtYc and CrtYd, members of the heterodimeric lycopene monocyclase family encoded by adjacent genes at *carB* [42]. The γ-carotene is subsequently hydroxylated by hydroxyneurosporene synthase (CrtC), desaturated by hydroxyneurosporene dehydrogenase (CrtD), and appended with a sugar moiety possibly by the action of the *orf6* gene product, a predicted glycosyltransferase. Finally, a putative acyltransferase encoded by *orf9* may act in myxobacton esterification [12]. Regulation of the carotenoid biosynthesis pathway can occur at the levels of transcription, post-transcription, modulation of enzyme activity through crosstalk and cooperation between them (such as the synergistic action of CrtIa and CrtIb mentioned above) and feedback regulation by the final carotenoid product and/or precursors. Moreover, how pathways intertwined with that for carotenogenesis, such as the MVA or the PPIX/heme biosynthesis pathways, are regulated can be important. Clearly, regulation at the transcription level is among the earliest and most crucial, and light-dependent regulation of transcription of *crtIb* and of the genes at *carB* has been intensely studied in *M. xanthus*.

5. Two Modes of Light Sensing and Signaling in *M. xanthus* Carotenogenesis

Since light triggers expression of *carB* and *crtIb*, understanding how light is sensed and converted to a cellular signal to mount the transcriptional response in *M. xanthus* is critical. In most living organisms including bacteria, the crucial task of sensing and transducing the light signal depends on photoreceptors, which are specialized proteins equipped with covalently or noncovalently bound light-sensing cofactors called chromophores. Photoreceptors have been classified into ten families thus far based on the specific chromophore and the protein photosensory domain [5,21,44–47]. One or more of these proteins occur in various bacteria, some are more widely distributed than others, and some occur even in species with minimal genomes or lifestyles that might suggest an absence of photoreceptors. Yet, surprisingly, given its well-established light response and a genome that is among the largest and most complex across bacteria, *M. xanthus* appeared to lack known photoreceptors. Consequently, blue light sensing through PPIX and the corresponding, rather convoluted, signaling pathway (discussed in Section 7) were considered as the sole mechanism to trigger carotenogenesis in *M. xanthus*. This changed about ten years ago with the discovery of the first member of an entirely new family of photoreceptors, which established a second light sensing and signaling mechanism that is simpler and more direct [20,21].

6. Direct Light Sensing, Signal Transduction and Gene Regulation by the B_{12}-Based CarH Photoreceptor

Discovery of the more direct light-sensing/signaling pathway and of a new photoreceptor family emerged from studies of the two adjacent and most downstream genes of the *carB* cluster, *carA* and *carH*, whose expression is significantly enhanced in the light [20,21,38,40]. Whereas mutations at *carA* yielded a Car^C phenotype, linking it to a negative regulator [32,34], a *carH* deletion had no apparent effect [48], even though the corresponding gene products of comparable sizes (CarA: 288 residues; CarH: 299 residues) share ~48% similarity (~35% sequence identity) and a similar two-domain architecture [38,48]. In both proteins, the ~70-residue N-terminal region resembles the DNA-binding domain (DBD) of MerR family proteins [38], which are widespread transcription factors in bacteria that repress or activate gene expression in response to diverse environmental stimuli such as oxidative stress, heavy metals or antibiotics [49]. MerR proteins bind as dimers via their winged-helix DNA binding domains to specific (pseudo)palindromic sites located within or overlapping their target primary σ^A-dependent promoters, and binding of a ligand (metal/drug) to a C-terminal module or oxidation of a redox center in it, enables these proteins to modulate transcription [49–52]. Notably, the ~200-residue C-terminal domain in CarA and CarH resembles a domain that binds to methylcobalamin (MeCbl) [48], one of the two biological forms of vitamin B_{12}, in the methionine biosynthesis enzyme MetH, a methionine synthase. The MetH B_{12}-binding domain (B_{12}-BD) houses a signature motif, $E/DxHx_2Gx_{41}SxTx_{22\text{-}27}GG$, whose His supplies the lower axial ligand in the so-called base-off/His-on binding to B_{12} [53]. Prior to CarA and CarH, such B_{12}-BDs were reported only in enzymes using B_{12} as a cofactor [54,55]. The combination of a B_{12}-BD and a DBD in CarA and CarH was therefore unprecedented and hinted at a pair of unusual transcription factor paralogs. Identifying a role for B_{12} and its mode of action, however, turned out to be less than straightforward.

6.1. CarH and Vitamin B_{12} Regulate Light-Induced Expression of Carotenoid Genes

Mapping the transcription start site at the *carB* locus identified a light-inducible primary σ^A-dependent promoter, P_B, with a consensus TTGACA –35 element and a less conserved TACCTC –10 element [38], which was recognized by σ^A-bound RNA polymerase (RNAP) in vitro [56]. CarA was found to dimerize via its B_{12}-BD [20,57] and use its N-terminal DBD (which indeed structurally resembles MerR DBDs [58]) to bind cooperatively, as two dimers, to a large ~55-bp DNA segment at the P_B promoter region (from positions –70 to –19 relative to the transcription start site) [56,57,59]. Since the operator overlaps with the –35 P_B promoter element, CarA binding can block promoter access to RNAP-σ^A and repress transcription [56]. Surprisingly, even though CarA could bind B_{12}, consistent with the presence of a canonical B_{12}-binding motif at its C-terminal domain, it neither required B_{12} for operator binding in vitro nor did mutating key residues in its B_{12}-binding motif impair P_B repression in the dark in vivo [60]. Key to unmasking the role of CarH was the finding that the Car^C phenotype caused by deleting *carA* could be reverted to wild-type behavior upon addition of exogenous vitamin B_{12} to the growth medium (*M. xanthus* takes up and assimilates B_{12} but cannot synthesize it de novo) [60]. CarH was shown to orchestrate this B_{12}-dependent repression of P_B in vivo and its relief in the light, and this activity of CarH required an intact CarA operator [60]. Thus, CarA and CarH both target the same operator at P_B to control light-induced expression of all but one of the carotenogenic genes in *M. xanthus*, but only CarH absolutely required B_{12} for activity. These findings not only established a functional link between B_{12} and CarH but also revealed a novel facet of this vitamin: its use in a cellular light response.

6.2. Molecular Architecture and Mode of Action of the B_{12}-Based CarH Photoreceptor

Answers to what specific B_{12} form was required by CarH and its molecular mechanism of action, as well as why and how CarH differs from its paralog CarA, began to emerge with a seminal study ten years ago [20]. CarA and CarH remained the first and only known transcription factors with a B_{12}-binding motif until homologs of unknown function were revealed in bacterial genomes covering a vast taxonomical range beyond myxobacteria [20,21]. This allowed comparative studies and better molecular understanding of these proteins. Whereas CarH has thus far resisted purification in the native form, two of its homologs from bacteria unrelated to *M. xanthus* have been purifiable in a native soluble form and could therefore be well-characterized in vitro. Both homologs turned out to be B_{12}-dependent like CarH. Studies of the homolog in the Gram-negative *Thermus thermophilus*, $CarH_{Tt}$, yielded valuable biochemical [20], structural [19] and photochemical insights [61–63] that were further extended with $CarH_{Bm}$, the homolog in the Gram-positive *Bacillus megaterium* [64,65]. These findings, reviewed elsewhere [21,22], are briefly highlighted here.

The specific B_{12} form required in CarH-mediated regulation of light-induced carotenogenesis in *M. xanthus* was established as 5′-deoxyadenosylcobalamin (AdoCbl) or coenzyme B_{12} (Figure 2a), which binds to the CarH C-terminal domain and directs its oligomerization and function [20]. AdoCbl is a complex organometallic molecule with a central cobalt, generally Co^{3+}/Co(III), coordinated to: (a) four equatorial pyrrolic nitrogens of the corrin ring; (b) a lower axial nitrogen from the 5,6-dimethylbenzimidazole (DMB) moiety linked to the corrin ring (so-called base-on or DMB-on conformation), or histidine from the B_{12}-binding motif in a protein (base-off/His-on binding, mentioned earlier); (c) an upper axial 5′-deoxyadenosyl (Ado) group; this upper ligand is methyl (Me) in MeCbl or cyano (CN) in vitamin B_{12}, a nonbiological form. The Co-C bond to an alkyl carbon in AdoCbl or MeCbl confers some unique and useful chemical properties. Its enzyme-catalyzed cleavage, which enables the use of AdoCbl in mutases, dehydratases, deaminases and ribonucleotide reductases and of MeCbl in methyltransferases, has been extensively studied and reviewed elsewhere [54,55]. Cleavage of the Co-C bond, by near-UV and visible light of wavelengths <530 nm, also underlies the use of AdoCbl as a chromophore for light sensing and response by CarH proteins (Figure 2b,c), which now represent a separate, large and widespread photoreceptor family among the ten currently known [19–22,61–63].

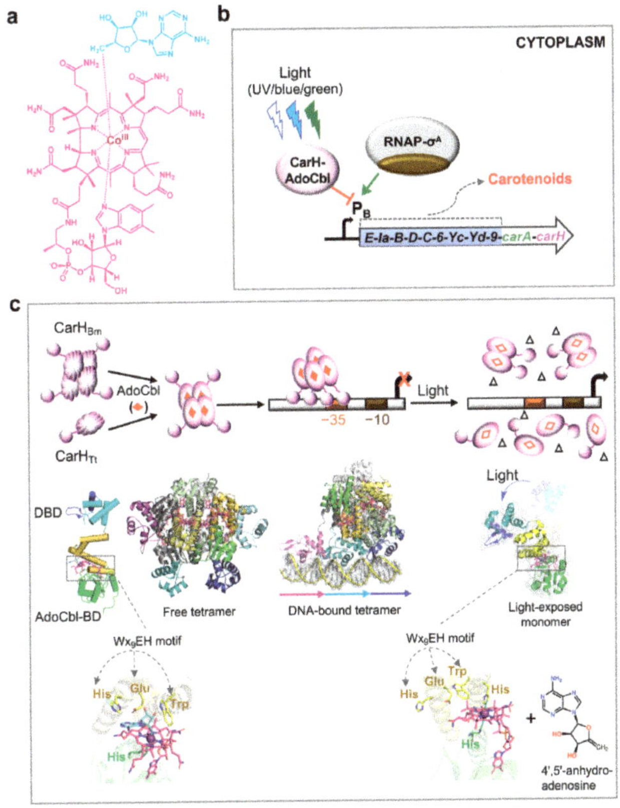

Figure 2. Light sensing and gene regulation by the B_{12}-based CarH photoreceptor. (**a**) Chemical structure of AdoCbl, the light-sensing chromophore of the CarH photoreceptor, with the upper axial 5′-deoxyadenosyl group in cyan and the rest of the molecule in magenta as depicted in the structures below. (**b**) CarH-mediated regulation at P_B. In the dark, AdoCbl-bound CarH binds to its operator at P_B to block access to RNAP-σ^A and repress transcription; and light (UV, blue or green) inactivates CarH to prevent its binding to operator, allowing P_B access to RNAP-σ^A and transcription initiation. (**c**) Molecular mechanism of CarH-mediated regulation at P_B. AdoCbl (filled red diamonds) bind to apo form monomers ($CarH_{Tt}$) or molten globule tetramers ($CarH_{Bm}$) to produce active, properly folded, compact tetramers that bind in the dark to an operator overlapping with a σ^A-dependent promoter (shown for –35 region but can be –10 or both) and thereby block transcription. UV, blue or green light photolyzes CarH-bound AdoCbl and disrupts DNA-bound tetramers to monomers ($CarH_{Tt}$) or dimers ($CarH_{Bm}$) that retain photolyzed AdoCbl (open red diamonds), leading to loss of operator binding and transcription. Upon photolysis, the upper ligand of AdoCbl is released as 4′-5′-anhydroadenosine (open triangles). Structures for AdoCbl-bound $CarH_{Tt}$ tetramer, free and DNA-bound, and for the light-exposed monomer are shown below. The protomer structure (left) with the DBD in cyan (recognition helix and wing, dark blue) and the AdoCbl-BD, with its four-helix bundle subdomain in golden, Rossmann fold subdomain in green, AdoCbl colored as in (**a**). PDB accession codes; 5C8D (tetramer in the dark), 5C8E (DNA-bound tetramer in the dark), 5C8F (light-exposed monomer). Below are close-ups of the Trp, Glu and His of the Wx_9EH motif capping the Ado group of AdoCbl in $CarH_{Tt}$, with the lower axial His (green) in the dark state (left) and light-exposed (right) state. In the latter, a His adjacent to the Trp in the Wx_9EH motif becomes the upper axial ligand in a bis-His linkage. The 4′-5′-anhydroadenosine product of AdoCbl-$CarH_{Tt}$ photolysis is also shown (far right bottom).

Studies of $CarH_{Tt}$ and $CarH_{Bm}$ revealed that light-dependent regulation of transcription relies on modulation of their oligomeric state by AdoCbl and light (Figure 2c) [19–22,64–66]. AdoCbl-free $apoCarH_{Tt}$ is a monomer and $apoCarH_{Bm}$ is a loosely folded molten globule tetramer, and both bind poorly to operator DNA. Both proteins form AdoCbl-bound tetramers in the dark that bind tightly to a large operator, which overlaps with the target gene promoter, to thwart access to RNAP-σ^A and block transcription; in the light, cleavage of the Co-C bond frees the upper axial Ado group and provokes tetramer disassembly to photolyzed $CarH_{Tt}$ monomers or $CarH_{Bm}$ dimers that detach from the operator to allow RNAP-σ^A binding and transcription initiation [19,20,64]. Cleavage of the AdoCbl chromophore with release of the Ado group is irreversible, in contrast to the usually reversible light-induced molecular changes observed with other photoreceptor chromophores [44,45,47], suggesting that there may be pathways to recover and reuse the chromophore that remain to be identified. Available data suggest that, like $CarH_{Tt}$, CarH is a monomer in the light-exposed AdoCbl-bound and apo forms, and that the dark AdoCbl-bound form is oligomeric but its stoichiometry remains to be defined [20].

Crystal structures of the AdoCbl-$CarH_{Tt}$ tetramer, free or DNA-bound, and of the light-exposed AdoCbl-$CarH_{Tt}$ monomer provided detailed molecular snapshots of CarH architecture and its light-dependent mechanism of action [19]. It confirmed the two-domain CarH modular architecture, with a MerR/CarA-like winged-helix N-terminal DBD connected by a flexible, disordered linker to a C-terminal AdoCbl-binding domain (hereafter AdoCbl-BD), in which AdoCbl is sandwiched between a four-helix bundle and a Rossmann fold subdomain (Figure 2c). The AdoCbl-BD is structurally similar to the MetH MeCbl-binding domain but has a critical Wx_9EH motif in the four-helix bundle that caps the upper axial Ado, which is absent in MetH (Figure 2c). In addition to the classic $ExHx_2Gx_{41}SxV/Tx_{22\text{-}27}GG$ B_{12}-binding motif, the Wx_9EH motif is absolutely conserved in all CarH homologs studied thus far and its critical role in AdoCbl-binding and function has been experimentally demonstrated [19,64]. Thus, absence of the motif in CarA can largely account for its B_{12}-independent activity. Indeed, the presence of both signature motifs defines CarH homologs, and several hundreds of these now assigned from genome data are broadly distributed across diverse bacterial taxa [21,22]. The dark state AdoCbl-$CarH_{Tt}$ tetramer is a dimer of two dimers, each of which is itself assembled by head-to-tail packing of two monomers via their AdoCbl-BD, with Trp of the Wx_9EH motif playing a crucial role. Since tetramer formation is very favorable, dimers are detected only by disruption of the dimer-dimer interface, such as by mutation [19,20,66]. In this unusual AdoCbl-$CarH_{Tt}$ tetramer assembly, the DBDs of neighbouring monomers point away from each other on the tetramer surface, which results in an unexpected DNA binding mode, wherein one DBD contacts a 11-bp direct repeat (DR) with a consensus nAnn**T**nn**ACA**n sequence (*n* = any base). Hence, it differs from the typical (pseudo)palindromic DNA sites of MerR proteins, yet it conserves most of the DNA contacts. Whereas three such tandem 11-bp DRs comprise the $CarH_{Tt}$ operator, four of these constitute the $CarH_{Bm}$ operator [19,64] and, likely, the ~55-bp *M. xanthus* CarH operator [20], suggesting a notable DNA-binding plasticity. Comparing the tetramer structure with that determined for the photolyzed $CarH_{Tt}$ monomer yielded molecular insights into light-induced tetramer collapse and loss of DNA binding. The light-exposed form revealed bound photolyzed AdoCbl (without the upper axial Ado) and a large shift (>8 Å) of the four-helix bundle relative to the Rossmann fold (Figure 2c), which disrupts the head-to-tail dimer interface, and thereby the tetramer, leading to loss of DNA binding.

The photochemistry of $CarH_{Tt}$-bound AdoCbl examined by analyzing photolysis products [61], by ultrafast spectroscopy [62,63] and by theoretical calculations [67] suggested it may differ significantly from that established for free or enzyme-bound AdoCbl. Photolytic cleavage of free AdoCbl, often a model for that in AdoCbl-dependent enzymes, is homolytic and generates reactive cob(II)alamin and Ado• radical species that rapidly react to yield specific products depending on the presence or otherwise of molecular oxygen [21,61,68]. In AdoCbl-dependent enzymes, which also rely on homolytic Co-C bond cleavage, the cob(II)alamin and Ado• radical species are generated in carefully controlled protein environments to ensure the difficult radical-based enzyme action and cofactor recovery, and to simultaneously limit enzyme damage and unwanted side reactions [69]. Since CarH controls a cell response (carotenogenesis) precisely to combat reactive ROS like $^1O_2$, its use of an AdoCbl chromophore with an underlying irreversible photolytic Co-C cleavage that releases reactive radicals seemed paradoxical. Remarkably, CarH appears to resolve this problem by altering AdoCbl photochemistry for its safe use as a photoreceptor chromophore. It was found that photolysis of $CarH_{Tt}$-bound AdoCbl avoids release of Ado• radicals by generating 4′,5′-anhydroadenosine, a harmless product undetected upon cleavage of free or enzyme-bound AdoCbl ([61]; Figure 2c). Based on ultrafast spectroscopy data, it has been proposed that CarH enables an unprecedented heterolytic cleavage of the AdoCbl Co–C bond to bypass radical formation and release [62] or stabilizes an excited state long enough to ensure the reactions that yield the 4′,5′-anhydroadenosine product [63]. The molecular mechanism for how CarH alters AdoCbl photochemistry is still unclear. It has been speculated that molecular oxygen and residues around the Ado group, notably of the Wx_9EH motif, may be important.

7. Blue Light Sensing, Signaling and Gene Regulation in the B_{12}-Independent Pathway

7.1. Light Is Perceived through Photoexcitation of PPIX, Which Leads to $^1O_2$ Production

Although the blue light-PPIX sensing and signaling mechanism to induce carotenogenesis in *M. xanthus* was the first to be identified, it is also the more complex one. Genetic evidence for the role of PPIX came from analysis of *M. xanthus* strains bearing specific deletions of genes in the heme biosynthetic pathway that resulted in elimination or overproduction of endogenous PPIX [30]. Thus, a strain with a deletion of *hemB*, which encodes an early enzyme in the heme biosynthetic pathway was Car^-, and the Car^+ phenotype could be restored by supplying PPIX exogenously. On the other hand, a strain with a deletion of *hemH*, whose product incorporates ferrous iron into PPIX in the final step of the heme biosynthetic pathway, exhibited a markedly enhanced light-induced carotenogenesis. The light response thus requires PPIX and correlates with the photosensitizer levels. The need for blue light and PPIX to induce carotenogenesis could be bypassed using the phenothiazinium dye methylene blue and red light, which also generates $^1O_2$, and was suppressed by $^1O_2$ quenchers [30]. The blue light signal is thus transduced via PPIX to $^1O_2$ and then relayed via a recently identified (and unprecedented) mechanism, whose molecular details continue to be unfurled.

7.2. CarF and Plasmalogen Lipids in M. xanthus Blue Light-PPIX-$^1O_2$ Signaling

Signaling by $^1O_2$ produced by blue-light photoexcitation of PPIX absolutely requires CarF [30], which was found in an analysis of Tn*5*-*lac* mutants and mapped to a locus unlinked to those previously identified in *M. xanthus* [70]. CarF is a 281-residue membrane protein with a four transmembrane-helix topology (Figure 3a), and its expression is not light-dependent [70,71]. Sequence homology searches [23] revealed that bacterial CarF-like proteins are present only in myxobacteria and a few Leptospiraceae and Alphaproteobacteria but, intriguingly, they are widespread in animals (invertebrates and vertebrates including humans, where the homolog is named TMEM189 or Kua [72]) and in plants. Protein phylogenetic analysis clearly indicated that CarF homologs from animals and from Leptospira are more related to those in *M. xanthus* and other myxobacteria, and those from Alphaproteobacteria and plants group together and are less related to CarF (Figure 3b). Until very recently their functions were largely unknown, except for the fact that CarF was required in the *M. xanthus* light response, and that a plant CarF homolog was a chloroplast fatty acid desaturase (FAD4) that generates an unusual trans double bond in the *sn*-2 acyl carbon chain [73].

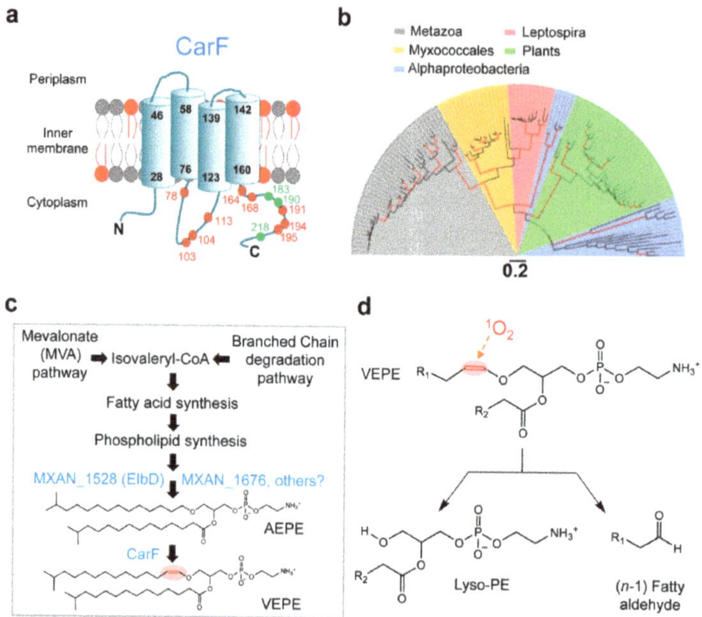

Figure 3. *M. xanthus* CarF and plasmalogen synthesis. (**a**) Cartoon representation of the *M. xanthus* CarF protein depicting its experimentally established membrane topology with four transmembrane helices (delimiting residues of each helix numbered in black). Numbered dots correspond to the 12 histidines in CarF. Nine of these (in red) are essential for CarF function but not the rest (in green); all nine essential histidines are conserved in animal CarF homologs, and all of these except His113 in plant homologs. The inner membrane plasmalogens are depicted in red. (**b**) Maximum-likelihood unrooted phylogenetic tree based on selected CarF homologs in metazoa, bacteria and plants (distributed in different colored sectors as indicated; branches in red, ≥75% confidence values from 200 bootstrap replicates; scale bar, number of substitutions per residue). (**c**) The *M. xanthus* plasmalogen biosynthesis pathway highlighting the early pathways and the final step, in which CarF mediates the desaturation that converts its alkyl ether lipid AEPE (1-*O*-(13-methyltetradecyl)-2-(13-methyltetradecanoyl)-glycero-3-phosphatidylethanolamine), to the plasmalogen VEPE (1-*O*-(13-methyl-1-*Z*-tetradecenyl)-2- (13-methyltetradecanoyl)-glycero-3-phosphatidylethanolamine). (**d**) Blue light-PPIX generated $^1O_2$ cleaves the vinyl ether bond of plasmalogens (VEPE) to yield a lyso-PE (2-monoacylglycerophosphoethanolamine) and an (*n-1*) fatty aldehyde (and formic acid, not shown).

A notable feature of CarF is its many (12) histidines, all cytoplasmic, with nine being essential for function ([23]; Figure 3a). The distribution of these histidines, some as HxxxH and HxxHH motifs, resembles that in membrane-associated diiron fatty acid desaturases and hydroxylases of otherwise low overall sequence similarity to CarF [23,70–72]. Hence, these observations hinted that CarF might be a fatty acid desaturase, like FAD4, but probably of a different kind, given that FAD4 lacks one of the crucial histidines in CarF [23].

The exact function of CarF and its role in *M. xanthus* light-induced carotenogenesis has only now been established [23]. It was discovered that CarF and its homologs in animals from worm and fly to fish, mouse and human, but not those in plants, correspond to the long-sought plasmanylethanolamine desaturase (now named PEDS1). This enzyme converts plasmanylethanolamine or alkyl ether phosphatidylethanolamine (glycerophospholipids with the *sn*-1 hydrocarbon chain linked by an ether bond instead of the typical ester bond; hereafter, AEPE) to plasmenylethanolamine, the alkenyl or vinyl ether phosphatidylethanolamine (hereafter, VEPE; Figure 3c). VEPE and analogs with choline instead of ethanolamine, collectively called plasmalogens, are found in animals and some anaerobic bacteria but not in plants, fungi or most aerobic bacteria except, notably, myxobacteria [23]. Human brain, heart and leukocytes are rich in plasmalogens, which occur in all subcellular membranes, and their deficiency or abnormal levels correlate with many disorders including cancer and Alzheimer's disease [74–76]. As a result of their vinyl ether bond, plasmalogens can affect membrane fluidity and function, and have a proposed antioxidant role given their sensitivity to cleavage by $^1O_2$ and other ROS [77]. However, plasmalogens had never been implicated in signaling photooxidative stress, a role that has now been clearly demonstrated in the *M. xanthus* light-induced carotenogenic response. Thus, deletion of *carF* annuls plasmalogen biosynthesis [23] as well as light-induced carotenogenesis [30,70], and the latter can be restored by supplying exogenous plasmalogens, even those from human cells that are distinct from the natural ones in *M. xanthus* (in that they have *sn*-1 and *sn*-2 moieties that differ from those in the *M. xanthus* VEPE). Furthermore, deleting genes (*elbD* and MXAN_1676) implicated in synthesis of the precursor AEPE (Figure 3c) impaired light-induced carotenogenesis, but was rescued by exogenous plasmalogen or by AEPE, which CarF converted to VEPE [23]. In sum, CarF is crucial in the response to light because it is indispensable for the biosynthesis of plasmalogens.

The role of plasmalogens in a blue light-PPIX-$^1O_2$ signaled response is both very recent and unprecedented, and identifying the underlying molecular mechanism of action is still being pursued. Breakage by $^1O_2$ of the vinyl ether bond in the plasmalogen yields lyso-PE (2-monoacylglycerophosphoethanolamine) and a fatty aldehyde (Figure 3d; [23,77–79]). This may perturb local membrane structure, environment and properties and affect the function(s) of downstream effector(s) in the pathway. The cleavage products might also function as signaling lipids or second messengers to modulate (or inactivate) effector activity through establishing noncovalent interactions, or covalent adducts between the reactive fatty aldehyde product and target nucleophiles (lysines, cysteines or histidines in proteins). Plasmalogens may themselves bind to specific membrane proteins or complexes to directly modulate their functions through interactions with $^1O_2$. These mechanisms, frequently invoked to link plasmalogens and cellular signaling [77,80], may also operate in *M. xanthus*.

7.3. Light-Induced Expression of the carQRS Operon and Gene crtIb

Early genetic analysis established that the *carR* locus encodes a negative regulator acting downstream of CarF [32,70], and that *carQ* and *carS*, closely linked to *carR*, encode positive regulators [33–37]. Subsequent DNA sequencing and transcription start site mapping revealed three translationally coupled genes, *carQ*, *carR* and *carS*, forming the *carQRS* operon and expressed from the light-inducible P_{QRS} promoter, which has –35 and –10 promoter elements divergent from typical *M. xanthus* RNAP-σ^A promoters [36]. Mutations at *carQ* are epistatic over those at *carR* and block activation of *carQRS* as well as of *crtIb*, the structural gene for carotenogenesis unlinked to the *carB* cluster [35–37,39]. Furthermore, *crtIb* expression is driven by a light-inducible promoter P_I, with –35 and –10 promoter elements similar to P_{QRS} [39,81]. These findings therefore implicated CarQ in activating *carQRS* and *crtIb* expression from similar light-dependent promoters, and CarR in their downregulation.

While CarS turned out to be the trans acting antirepressor of CarA [57,82], CarQ was identified as the founding member of a new, large and diverse group of alternative σ factors known as the extracytoplasmic function or ECF-σ factors, which were first discovered over 25 years ago [83,84]. Usually, ECF-σ act in a gamut of cellular responses to a variety of extracytoplasmic stimuli (hence the name) and are negatively regulated by association with cognate anti-σ factors, which are often membrane-bound and coexpressed with their ECF-σ partner [85]. CarR was shown to be such a membrane-bound anti-σ, as it specifically and stoichiometrically sequestered CarQ and rendered it inactive in the dark [37] through direct, physical interactions ([71,86]; Figure 4). With six transmembrane helices [36,86,87],

CarR belongs to a small group of anti-σ factors with similar membrane topology, largely restricted to proteobacteria, and classified as DUF1109 in the conserved protein domain family database [85]. Some of these other anti-σ act in stress responses to ROS or to heavy metals [88–90] and, interestingly, transcription of both *carQRS* and *crtIb* is activated in the dark by copper [91]. The molecular basis for this copper-mediated action, which bypasses both CarF and light, is still unknown and remains to be elucidated.

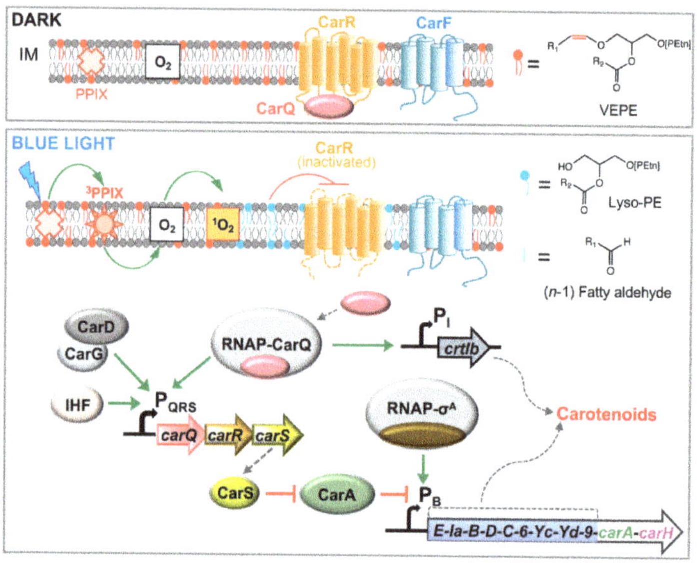

Figure 4. Model for the blue light-PPIX-$^1O_2$ signaling and transduction pathway, and regulation by the CarA-CarS repressor-antirepressor pair in *M. xanthus*. In the dark, the anti-σ factor CarR (with a six transmembrane-helix topology; IM: inner membrane) sequesters its cognate ECF-σ factor CarQ. Blue light excites PPIX to the high-energy ^{3}PPIX state, from which energy transfer to molecular O_2 generates the highly reactive $^1O_2$. CarF produces plasmalogens (VEPE), which are required to transmit the $^1O_2$ signal and cause the inactivation of CarR by a mechanism that remains to be elucidated. Plasmalogen cleavage by $^1O_2$ might perturb the local membrane environment of CarR, or its cleavage products may interact with CarR, to alter its activity. This liberates CarQ, which associates with RNAP to activate promoters P_{QRS} (which also requires the CarD-CarG global regulatory complex and IHF) and P_I to drive expression of the regulatory *carQRS* operon and of the carotenogenic *crtIb* gene, respectively. CarS, expressed from P_{QRS} in the light, counteracts repression of P_B by CarA (see below) to drive expression of the *carB* operon, containing all but one of the carotenogenic genes, leading to the synthesis of carotenoids.

CarQ must be first liberated from its cognate anti-σ CarR, which sequesters it in the dark [37], to associate with RNAP and initiate transcription of its target genes ([86]; Figure 4). Light triggers this liberation of CarQ from CarR, but the exact molecular mechanism remains elusive. CarR was reportedly unstable when exposed to light, especially when cells enter the stationary phase of growth [86], which also correlates with PPIX accumulation. Increased PPIX levels, however, do not activate P_{QRS} and *carQRS* expression in the absence of CarF [30], and since the actual role of CarF is in plasmalogen synthesis, this lipid must somehow mediate inactivation of CarR by light [23]. Various mechanisms, as noted before, can be hypothesized for how plasmalogens mediate CarR activation. Plasmalogen cleavage by $^1O_2$ might perturb the local membrane environment of CarR, or a cleavage product may interact with CarR to alter its activity. Other unknown player(s) or mechanism(s) cannot also be ruled out. These questions will have to be resolved in future work.

7.4. Regulation of CarQ Activity in Light-Induced Expression of carQRS and crtIb

Negative regulation by CarR is the key determinant of CarQ activity since this controls its availability for association with RNAP. Nonetheless, additional factors required for CarQ activity have also been discovered. An early screen of Tn*5-lac* Car^{-} mutants identified two constitutively expressed genes, *carD* and *ihfA*, acting directly in light-induced activation of *carQRS* and, through expression of CarQ and CarS, indirectly in those of *crtIb* and the *carB* operon, respectively [92,93]. The *ihfA* gene encodes the α subunit of the integration host factor (IHF) heterodimer, a nucleoid-associated, histone-like architectural factor that functions as a global regulator [94,95]. Gene *carD* encodes a 316-residue DNA-binding transcriptional factor and is translationally coupled to a downstream gene, *carG*, whose product forms with CarD a tight heteromeric complex that functions as one regulatory unit ([96–98]; Figure 4). CarG is therefore essential for CarD function and the two always coexist. Interestingly, the pair occurs exclusively in *M. xanthus* and related myxobacteria. Thus, at least three other proteins besides CarR, namely IHF, CarD and CarG, regulate CarQ activity at P_{QRS}.

Both CarD and CarG are unusual transcription factors. CarG is a monomer with no DNA-binding capacity, which coordinates two zinc atoms via a His-Cys rich segment ($HQx_2Hx_2Ex_2HCx_4CxMx_{16}Cx_2C$; x is any amino acid) [96]. The motif is similar to one found in zinc-metalloproteases called metzincins [99] but an E essential for protease activity is replaced by Q in the motif in CarG, which has no protease activity [96]. In short, CarG can be considered to be one more among the few bacterial transcriptional factors that do not bind DNA [100,101], but which appears exclusively in myxobacteria. CarD is also a rather singular protein. One striking feature is its ~136-residue C-terminal segment comprising a highly acidic ~50-residue region flanked by a C-terminal segment containing four repeats of the RGRP "AT-hook" DNA-binding motif (Figure 5; [102]). Interestingly, these motifs are rare in bacteria but occur in eukaryotic proteins such as high-mobility group type A (HMGA), a relatively abundant, nonhistone architectural factor that remodels chromatin in various DNA transactions [97,102–104]. Similar to HMGA, the CarD C-terminal domain is intrinsically disordered and binds to the minor groove of appropriately spaced AT-rich DNA tracts; and two such tracts at P_{QRS} (at –63 and –77 relative to the transcription start site) to which CarD binds are implicated in CarQ activity [96,98,103,105]. In line with this, the minimum P_{QRS} segment required for CarQ activity is a ~145 bp upstream stretch starting from the transcription start site [106]. By comparison, CarQ activity at its other target promoter, P_I, which CarD and IHF affect indirectly, requires a shorter stretch extending to position –54 upstream of the transcription start site [81]. Interestingly, CarD can function in *M. xanthus* even when its natural HMGA-like domain is replaced by human HMGA, histone H1 or the intrinsically disordered H1 C-terminal region, indicating that a basic, structurally disordered C-terminal domain is sufficient for CarD function [98]. Surprisingly, even without its HMGA-like domain, CarD functions in vivo, albeit with diminished activity [107]. By contrast, the remaining N-terminal region of CarD is indispensable for function [108].

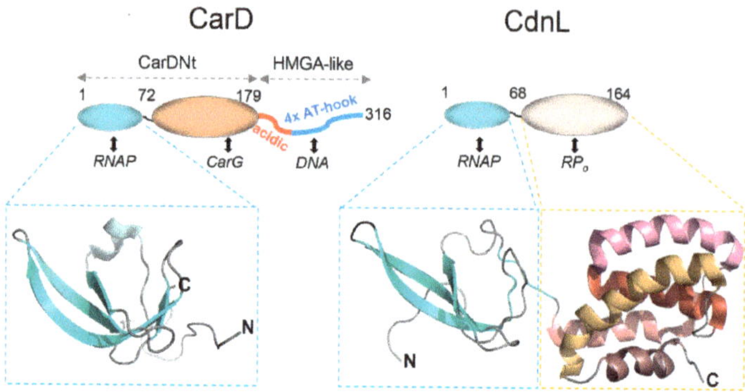

Figure 5. *M. xanthus* CarD and CdnL. Schematics summarizing structural and functional domains of *M. xanthus* CarD (left) and CdnL (right). Numbers correspond to the residues delimiting the indicated domains, which interact with the partners listed below. Bottom: Structures determined for the RNAP interacting module of CarD (PDB accession code 2LT1) and full-length CdnL (PDB accession code 2LWJ).

Unlike its intrinsically disordered HMGA-like C-terminal domain, the CarD N-terminal domain, CarDNt (Figure 5), is a structurally defined module with sequence similarity to the RNAP-binding domain of bacterial transcription repair coupling factors or TRCFs [103], which repair lesions in the transcribed strand by interacting with RNAP [109]. Indeed, CarDNt has an N-terminal subdomain with a five-stranded β-sheet Tudor-like tertiary structure (Figure 5) similar to its counterpart in TRCF [107] and interacts specifically with the RNAP β subunit [105,110]. Moreover, the C-terminal part of CarDNt, not involved in the interaction with RNAP, binds to CarG [96,98,107]. CarDNt is thus a protein–protein interaction hub directing interactions with both RNAP and CarG, while the CarD HMGA-like C-terminal domain mediates DNA binding.

Remarkably, although CarD homologs are restricted to myxobacteria closely related to *M. xanthus*, CarDNt is a defining member of a large family of bacterial RNAP-interacting proteins (PF02559 or CarD_CdnL_TRCF protein family; http://pfam.sanger.ac.uk, accessed on 29 April 2021) that includes not only CarD and TRCF homologs but also a large group of standalone proteins similar to CarD without its HMGA-like domain. These proteins, denoted CdnL (for CarD N-terminal Like), are widely distributed in bacteria and occur in *M. xanthus* and other δ-proteobacteria, α-proteobacteria, Actinomycetes, Firmicutes, Deinococcus-Thermus and Spirochaetes, but not in β-, γ- or ε-proteobacteria, Chlamydiae or Cyanobacteria [108,110,111]. Whereas knocking out *carD* does not affect normal growth or viability, CdnL is indispensable for normal growth and survival of *M. xanthus* [110,112,113]. CdnL has also been reported to be an essential gene in *Borrelia burgdorferi* (spirochaetes), *Mycobacterium tuberculosis* (mycobacteria) and *Rhodobacter sphaeroides* (α-proteobacteria), and to impair normal growth in *Caulobacter crescentus* (α-proteobacteria) [111,114–116]. This is because transcription of the essential RNAP-σ^A-dependent rRNA genes in these bacteria requires CdnL in the critical step of open promoter complex (RP_o) formation, and CdnL directly or indirectly impacts expression of important biosynthetic genes [111,113,116–118]. In contrast to CarD, which interacts with both RNAP and CarG via CarDNt [105,107], CdnL interacts only with RNAP [110] and does not bind DNA, since it lacks the HMGA-like domain [113]. Nevertheless, the CdnL N-terminal region conserves both the structure and contacts with the RNAP β subunit of its equivalent in CarDNt (Figure 5; [107,113]). When associated to RNAP in RP_o, the compact C-terminal region of CdnL can interact with promoter DNA from positions −14 to −10 to stabilize the transcription bubble [101,117]. Some of the functionally important residues in this CdnL domain, which comprises five well-packed α-helices (Figure 5; [113]), are conserved and important for CarD function as well, and mutating these affects CarD function at target promoters even though binding to CarG remains unaffected [107].

Importantly, the CarD-CarG complex affects processes other than light-induced carotenogenesis in *M. xanthus*. It was implicated in regulating the expression of some early genes in the starvation-induced development to multicellular fruiting bodies [92,96] and of various vegetatively expressed genes of mostly unknown functions [119], none CarQ-dependent. A later study showed that the CarD-CarG complex affects the activities of at least twelve ECF-σ/anti-σ pairs besides CarQ-CarR in *M. xanthus*, suggesting that the complex may control many of the ~45 putative ECF-σ factors in this bacterium [87]. Except for the light-induced CarQ-CarR pair, the signals that activate each of the other ECF-σ/anti-σ pairs that depend on CarD-CarG are unknown. One pair was, however, recently shown to direct the expression of one of the three CRISPR-Cas systems (type III-B) in *M. xanthus*, which suggested that this bacterial defense system is triggered by a phage [120]. Thus, CarD-CarG is a global regulator, like CdnL, but targets different genes. The parallels with CdnL (despite differences) and the finding that the CarD-CarG complex targets various ECF-σ promoters suggests that CarD-CarG may have a role at these promoters analogous to that of CdnL at RNAP-σ^A-dependent promoters [107]. This remains to be further explored in future studies.

7.5. Derepression of P_B by Light-Induced Expression of the CarS Antirepressor

The photoregulatory switch controlling expression of the *carB* cluster from the P_B promoter relies on repression by both CarH and CarA, and their inactivation by light. Whereas CarH is a photoreceptor that directly senses light, CarA repression is relieved by physical interaction with the CarS antirepressor, whose expression is induced by light (Figure 6; [56–59,82,121]). A series of biochemical, structural and mutational studies demonstrated that the CarA N-terminal domain is an autonomous folding unit with the winged-helix topology of MerR family DBDs (Figure 6), and that it contains the determinants for specific binding to operator DNA as well as to CarS [57–59]. Further structural-mutational analysis revealed that the highly acidic, 111-residue CarS adopts a five-stranded, antiparallel β-sheet fold resembling SH3 domains (protein–protein interaction modules prevalent in eukaryotes but rare in prokaryotes) and contains a solvent-exposed hydrophobic pocket lined by acidic residues that mimics operator DNA to bind tightly to the DNA recognition helix of CarA and sequester it (Figure 6; [121]). Interestingly, a gain-of-function *carS* mutant (*carS1*) lacking the 25 C-terminal residues results in constitutive, light-independent expression at P_B [36], presumably because the variant CarS1 is more acidic than CarS and thus binds more tightly to CarA [57]. Given that CarH recognizes the same operator as CarA and both proteins have similar DBDs and recognition helices, CarH also physically interacts with CarS, albeit with lower affinity than CarA [20,60,121]. Thus, repression of P_B by CarA is counteracted by CarS expressed only under light, while P_B repression by CarH is relieved mostly by the direct effect of light on the AdoCbl chromophore. CarS homologs occur only in myxobacteria related to *M. xanthus*, and likely play an analogous antirepressor role.

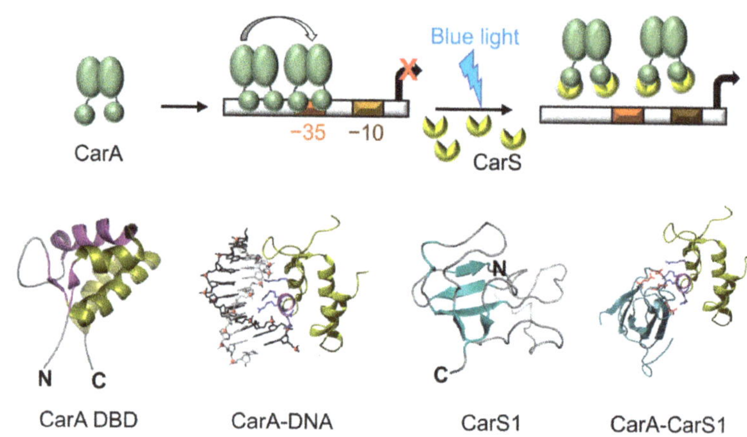

Figure 6. Molecular mechanism of CarA-CarS repressor-antirepressor mode of regulation at P_B. In the dark, CarA dimers bind cooperatively to its operator at P_B, which blocks access to RNAP-σ^A and represses transcription. CarS, expressed from the *carQRS* operon in the light, acts as a DNA mimic to sequester the CarA DBD and prevent its binding to operator, thereby enabling transcription initiation by RNAP-σ^A at P_B. Bottom: structures of the CarA DBD and CarS1 (PDB accession codes 2JML and 2KSS, respectively) and structural models for CarA-DNA and CarA-CarS1 complexes.

8. Conclusions

Delving into how *M. xanthus* "sees" and mounts a photooxidative stress response that triggers carotenogenesis uncovered two novel pathways in bacterial light sensing, signal transduction and gene regulation. One pathway relies on a form of vitamin B_{12} and its association with a single photoreceptor-cum-transcriptional factor, and the other is a B_{12}-independent, more complex route that requires various singular factors. Many worthy firsts can be credited to elucidation of the two pathways, including the discovery of one of the first ECF-σ factors, CarQ [83,84]; the founding members of large protein families, notably the B_{12}-based CarH photoreceptor family [19–22] and the CarD_CdnL family of RNAP-binding transcription factors [102,103,110]; the long-sought human desaturase involved in plasmalogen biosynthesis through its *M. xanthus* CarF homolog [23]. Insights specific to *M. xanthus* and closely related bacteria, but also ones more broadly conserved across bacteria, have emerged. This photooxidative stress response is linked, directly or indirectly, to that of copper and to heme and fatty acid biosynthesis, and shares global regulators with processes as diverse as fruiting body development and activation of CRISPR-Cas systems. Future work will undoubtedly reveal new, possibly surprising, interconnections to other cellular activities.

Beyond bacterial physiology, signaling and gene regulation, the findings from *M. xanthus* light-induced carotenogenesis have had other important ramifications. How this response and its unique factors are conserved across bacteria and other organisms provides valuable evolutionary insights. Some of the factors involved, which are more typical of eukaryotes, yield phylogenetic signals that may be supportive of the hypothesis that an ancient myxobacterium may have contributed in eukaryogenesis [122]. This hypothesis, known as the Syntrophy hypothesis for the origin of eukaryotes, posits that the eukaryotic cell evolved from symbiosis or syntrophy between a complex early myxobacterial-like deltaproteobacterium (host), an endosymbiotic Asgard-like archaeon (future nucleus) and an alphaproteobacterium (future mitochondrion) [122]. The role of a myxobacterium proposed in this hypothesis was based on the many myxobacterial-like genes in eukaryotes. These phylogenetic signals include, among various others, isoprenoid biosynthesis enzymes, HMGA proteins (CarD) [122], CarF and plasmalogens [23].

Satisfyingly, CarH has now been exploited as one of the few green-light responsive optogenetic tools for light-controlled: (a) gene expression in *M. xanthus* and transgene expression in mammalian and plant cells; (b) receptor interactions and signaling in human cells and zebra fish embryos; (c) generation of protein hydrogels that enable facile encapsulation and release of cells and proteins, and cell adhesions [123–128]. Notably, this last application was very recently adapted to address challenges in regenerative neurobiology to engineer metal-coordinated protein hydrogels for sustained delivery of neuroprotective cytokines aimed at neuronal survival and axon regeneration in vivo [126].

The discovery that CarF and its human and animal homologs are identical lipid desaturases essential in plasmalogen synthesis has not only revealed a remarkable conservation of this enzyme across a vast evolutionary distance, but also has important implications in human health and disease [23]. Plasmalogens have been linked to various human disorders including cancer and Alhzeimer´s disease but the unknown identity of plasmanylethanolamine desaturase had been an impediment in directly assessing the role of these lipids in diverse pathologies. This is now possible with the identity of the enzyme in hand, and has already proved useful in studies of mitochondrial metabolism [129] and ferroptosis [130,131].

Author Contributions: Conceptualization, writing—original draft preparation, supervision and funding acquisition S.P. and M.E.-A.; writing—review and editing, all authors. All authors have contributed to the work, read and agreed to the published version of the manuscript.

Funding: This research was funded by grants PGC2018-094635-B-C21 (to M.E.-A.) and PGC2018-094635-B-C22 (to S.P) from the Agencia Estatal de Investigación (AEI)-Spain and European Regional Development Fund (FEDER), and by grant 20992/PI/18 (to M.E.-A.) from Fundación Séneca (Murcia)-Spain. The Ministerio de Educación y Cultura-Spain funded Ph.D. fellowships to A.J.M.-G, E.P.-M. and E.B.-M., and AEI-Spain funded that to R.P.-C.

Data Availability Statement: Not applicable.

Acknowledgments: We thank Francisco J. Murillo, all past and present members of our group, and our various research collaborators for their valuable contributions.

Conflicts of Interest: The authors declare no conflict of interest.

References

1. Croce, R.; van Amerongen, H. Natural strategies for photosynthetic light harvesting. *Nat. Chem. Biol.* **2014**, *10*, 492–501. [CrossRef] [PubMed]
2. Croce, R.; van Amerongen, H. Light harvesting in oxygenic photosynthesis: Structural biology meets spectroscopy. *Science* **2020**, *369*, eaay2058. [CrossRef] [PubMed]
3. Saer, R.G.; Blankenship, R.E. Light harvesting in phototrophic bacteria: Structure and function. *Biochem. J.* **2017**, *474*, 2107–2131. [CrossRef]
4. Elías-Arnanz, M.; Padmanabhan, S.; Murillo, F.J. Light-dependent gene regulation in nonphototrophic bacteria. *Curr. Opin. Microbiol.* **2011**, *14*, 128–135. [CrossRef]
5. Purcell, E.B.; Crosson, S. Photoregulation in prokaryotes. *Curr. Opin. Microbiol.* **2008**, *11*, 168–178. [CrossRef]
6. Cohen, S.E.; Golden, S.S. Circadian rhythms in cyanobacteria. *Microbiol. Mol. Biol. Rev.* **2015**, *79*, 373–385. [CrossRef] [PubMed]
7. Zheng, E.-C.; Bosman, J.; Sartor, F.; Dodd, A.N.; Kovács, A.T.; Merrow, M. A circadian clock in a nonphotosynthetic prokaryote. *Sci. Adv.* **2021**, *7*, eabe2086.
8. Glaeser, J.; Nuss, A.M.; Berghoff, B.A.; Klug, G. Singlet oxygen stress in microorganisms. *Adv. Microb. Physiol.* **2011**, *58*, 141–173.
9. Ziegelhoffer, E.C.; Donohue, T.J. Bacterial responses to photo-oxidative stress. *Nat. Rev. Microbiol.* **2009**, *7*, 856–863. [CrossRef] [PubMed]
10. Di Mascio, P.; Martinez, G.R.; Miyamoto, S.; Ronsein, G.E.; Medeiros, M.H.G.; Cadet, J. Singlet molecular oxygen reactions with nucleic acids, lipids, and proteins. *Chem. Rev.* **2019**, *119*, 2043–2086. [CrossRef]
11. Sies, H.; Jones, D.P. Reactive oxygen species (ROS) as pleiotropic physiological signalling agents. *Nat. Rev. Mol. Cell Biol.* **2020**, *21*, 363–383. [CrossRef]
12. Elías-Arnanz, M.; Fontes, M.; Padmanabhan, S. Carotenogenesis in *Myxococcus xanthus*: A complex regulatory network. In *Myxobacteria: Multicellularity and Differentiation*; Whitworth, D.E., Ed.; ASM Press: Washington, DC, USA, 2008; pp. 211–225. [CrossRef]

13. Rodriguez-Concepción, M.; Avalos, J.; Bonet, M.L.; Boronat, A.; Gómez-Gómez, L.; Hornero-Mendez, D.; Limon, M.C.; Meléndez-Martínez, A.J.; Olmedilla-Alonso, B.; Palou, A.; et al. A global perspective on carotenoids: Metabolism, biotechnology, and benefits for nutrition and health. *Prog. Lipid Res.* **2018**, *70*, 62–93. [CrossRef]
14. Moise, A.R.; Al-Babili, S.; Wurtzel, E.T. Mechanistic aspects of carotenoid biosynthesis. *Chem. Rev.* **2014**, *114*, 164–193. [CrossRef]
15. Llorente, B.; Martinez-Garcia, J.F.; Stange, C.; Rodriguez-Concepcion, M. Illuminating colors: Regulation of carotenoid biosynthesis and accumulation by light. *Curr. Opin. Plant Biol.* **2017**, *37*, 49–55. [CrossRef]
16. Stanley, L.; Yuan, Y.W. Transcriptional regulation of carotenoid biosynthesis in plants: So many regulators, so little consensus. *Front. Plant Sci.* **2019**, *10*, 1017. [CrossRef]
17. Corrochano, L.M. Light in the fungal world: From photoreception to gene transcription and beyond. *Annu. Rev. Genet.* **2019**, *53*, 149–170. [CrossRef] [PubMed]
18. Yu, Z.; Fischer, R. Light sensing and responses in fungi. *Nat. Rev. Microbiol.* **2019**, *17*, 25–36. [CrossRef] [PubMed]
19. Jost, M.; Fernández-Zapata, J.; Polanco, M.C.; Ortiz-Guerrero, J.M.; Chen, P.Y.; Kang, G.; Padmanabhan, S.; Elías-Arnanz, M.; Drennan, C.L. Structural basis for gene regulation by a B_{12}-dependent photoreceptor. *Nature* **2015**, *526*, 536–541. [CrossRef]
20. Ortiz-Guerrero, J.M.; Polanco, M.C.; Murillo, F.J.; Padmanabhan, S.; Elías-Arnanz, M. Light-dependent gene regulation by a coenzyme B_{12}-based photoreceptor. *Proc. Natl. Acad. Sci. USA* **2011**, *108*, 7565–7570. [CrossRef]
21. Padmanabhan, S.; Jost, M.; Drennan, C.L.; Elías-Arnanz, M. A new facet of vitamin B_{12}: Gene regulation by cobalamin-based photoreceptors. *Annu. Rev. Biochem.* **2017**, *86*, 485–514. [CrossRef] [PubMed]
22. Padmanabhan, S.; Pérez-Castaño, R.; Elías-Arnanz, M. B_{12}-based photoreceptors: From structure and function to applications in optogenetics and synthetic biology. *Curr. Opin. Struct. Biol.* **2019**, *57*, 47–55. [CrossRef]
23. Gallego-García, A.; Monera-Girona, A.J.; Pajares-Martínez, E.; Bastida-Martínez, E.; Pérez-Castaño, R.; Iniesta, A.A.; Fontes, M.; Padmanabhan, S.; Elías-Arnanz, M. A bacterial light response reveals an orphan desaturase for human plasmalogen synthesis. *Science* **2019**, *366*, 128–132. [CrossRef]
24. Hoshino, Y.; Gaucher, E.A. On the origin of isoprenoid biosynthesis. *Mol. Biol. Evol.* **2018**, *35*, 2185–2197. [CrossRef]
25. Meiser, P.; Bode, H.B.; Muller, R. The unique DKxanthene secondary metabolite family from the myxobacterium *Myxococcus xanthus* is required for developmental sporulation. *Proc. Natl. Acad. Sci. USA* **2006**, *103*, 19128–19133. [CrossRef] [PubMed]
26. Burchard, R.P.; Dworkin, M. Light-induced lysis and carotenogenesis in *Myxococcus xanthus*. *J. Bacteriol.* **1966**, *91*, 535–545. [CrossRef] [PubMed]
27. Burchard, R.P.; Hendricks, S.B. Action spectrum for carotenogenesis in *Myxococcus xanthus*. *J. Bacteriol.* **1969**, *97*, 1165–1168. [CrossRef] [PubMed]
28. Ogilby, P.R. Singlet oxygen: There is indeed something new under the sun. *Chem. Soc. Rev.* **2010**, *39*, 3181–3209. [CrossRef] [PubMed]
29. Hodgson, D.A.; Murillo, F.J. Genetics of regulation and pathway of synthesis of carotenoids. In *Myxobacteria II*; Dworkin, M., Kaiser, D., Eds.; American Society for Microbiology: Washington, DC, USA, 1993; pp. 157–181.
30. Galbis-Martínez, M.; Padmanabhan, S.; Murillo, F.J.; Elías-Arnanz, M. CarF mediates signaling by singlet oxygen, generated via photoexcited protoporphyrin IX, in *Myxococcus xanthus* light-induced carotenogenesis. *J. Bacteriol.* **2012**, *194*, 1427–1436. [CrossRef] [PubMed]
31. Martínez-Laborda, A.; Elías, M.; Ruiz-Vázquez, R.; Murillo, F.J. Insertions of Tn*5* linked to mutations affecting carotenoid synthesis in *Myxococcus xanthus*. *Mol. Gen. Genet.* **1986**, *205*, 107–114. [CrossRef]
32. Balsalobre, J.M.; Ruiz-Vázquez, R.M.; Murillo, F.J. Light induction of gene expression in *Myxococcus xanthus*. *Proc. Natl. Acad. Sci. USA* **1987**, *84*, 2359–2362. [CrossRef] [PubMed]
33. Martínez-Laborda, A.; Balsalobre, J.M.; Fontes, M.; Murillo, F.J. Accumulation of carotenoids in structural and regulatory mutants of the bacterium *Myxococcus xanthus*. *Mol. Gen. Genet.* **1990**, *223*, 205–210. [CrossRef]
34. Martinez-Laborda, A.; Murillo, F.J. Genic and allelic interactions in the carotenogenic response of *Myxococcus xanthus* to blue light. *Genetics* **1989**, *122*, 481–490. [CrossRef]
35. Hodgson, D.A. Light-induced carotenogenesis in *Myxococcus xanthus*: Genetic analysis of the *carR* region. *Mol. Microbiol.* **1993**, *7*, 471–488. [CrossRef] [PubMed]
36. McGowan, S.J.; Gorham, H.C.; Hodgson, D.A. Light-induced carotenogenesis in *Myxococcus xanthus*: DNA sequence analysis of the *carR* region. *Mol. Microbiol.* **1993**, *10*, 713–735. [CrossRef] [PubMed]
37. Gorham, H.C.; McGowan, S.J.; Robson, P.R.; Hodgson, D.A. Light-induced carotenogenesis in *Myxococcus xanthus*: Light-dependent membrane sequestration of ECF sigma factor CarQ by anti-sigma factor CarR. *Mol. Microbiol.* **1996**, *19*, 171–186. [CrossRef] [PubMed]
38. Botella, J.A.; Murillo, F.J.; Ruiz-Vazquez, R. A cluster of structural and regulatory genes for light-induced carotenogenesis in *Myxococcus xanthus*. *Eur. J. Biochem.* **1995**, *233*, 238–248. [CrossRef]
39. Fontes, M.; Ruiz-Vázquez, R.; Murillo, F.J. Growth phase dependence of the activation of a bacterial gene for carotenoid synthesis by blue light. *EMBO J.* **1993**, *12*, 1265–1275. [CrossRef] [PubMed]
40. Ruiz-Vázquez, R.; Fontes, M.; Murillo, F.J. Clustering and co-ordinated activation of carotenoid genes in *Myxococcus xanthus* by blue light. *Mol. Microbiol.* **1993**, *10*, 25–34. [CrossRef] [PubMed]
41. Iniesta, A.A.; Cervantes, M.; Murillo, F.J. Cooperation of two carotene desaturases in the production of lycopene in *Myxococcus xanthus*. *FEBS J.* **2007**, *274*, 4306–4314. [CrossRef]

42. Iniesta, A.A.; Cervantes, M.; Murillo, F.J. Conversion of the lycopene monocyclase of *Myxococcus xanthus* into a bicyclase. *Appl. Microbiol. Biotechnol.* **2008**, *79*, 793–802. [CrossRef]
43. Reichenbach, H.; Kleinig, H. Pigments of myxobacteria. In *Myxobacteria: Development and Cell Interactions*; Rosenberg, E., Ed.; Springer: New York, NY, USA, 1984; pp. 128–137. [CrossRef]
44. Losi, A.; Gardner, K.H.; Möglich, A. Blue-light receptors for optogenetics. *Chem. Rev.* **2018**, *118*, 10659–10709. [CrossRef] [PubMed]
45. Möglich, A.; Yang, X.; Ayers, R.A.; Moffat, K. Structure and function of plant photoreceptors. *Annu. Rev. Plant Biol.* **2010**, *61*, 21–47. [CrossRef] [PubMed]
46. Gomelsky, M.; Hoff, W.D. Light helps bacteria make important lifestyle decisions. *Trends Microbiol.* **2011**, *19*, 441–448. [CrossRef] [PubMed]
47. Seong, J.; Lin, M.Z. Optobiochemistry: Genetically encoded control of protein activity by light. *Annu. Rev. Biochem.* **2021**. [CrossRef]
48. Cervantes, M.; Murillo, F.J. Role for vitamin B_{12} in light induction of gene expression in the bacterium *Myxococcus xanthus*. *J. Bacteriol.* **2002**, *184*, 2215–2224. [CrossRef]
49. Brown, N.L.; Stoyanov, J.V.; Kidd, S.P.; Hobman, J.L. The MerR family of transcriptional regulators. *FEMS Microbiol. Rev.* **2003**, *27*, 145–163. [CrossRef]
50. Chang, C.C.; Lin, L.Y.; Zou, X.W.; Huang, C.C.; Chan, N.L. Structural basis of the mercury(II)-mediated conformational switching of the dual-function transcriptional regulator MerR. *Nucleic Acids Res.* **2015**, *43*, 7612–7623. [CrossRef]
51. Fang, C.; Philips, S.J.; Wu, X.; Chen, K.; Shi, J.; Shen, L.; Xu, J.; Feng, Y.; O'Halloran, T.V.; Zhang, Y. CueR activates transcription through a DNA distortion mechanism. *Nat. Chem. Biol.* **2021**, *17*, 57–64. [CrossRef]
52. Philips, S.J.; Canalizo-Hernandez, M.; Yildirim, I.; Schatz, G.C.; Mondragón, A.; O'Halloran, T.V. TRANSCRIPTION. Allosteric transcriptional regulation via changes in the overall topology of the core promoter. *Science* **2015**, *349*, 877–881. [CrossRef] [PubMed]
53. Drennan, C.L.; Huang, S.; Drummond, J.T.; Matthews, R.G.; Ludwig, M.L. How a protein binds B_{12}: A 3.0 A X-ray structure of B_{12}-binding domains of methionine synthase. *Science* **1994**, *266*, 1669–1674. [CrossRef]
54. Banerjee, R.; Ragsdale, S.W. The many faces of vitamin B_{12}: Catalysis by cobalamin-dependent enzymes. *Annu. Rev. Biochem.* **2003**, *72*, 209–247. [CrossRef]
55. Giedyk, M.; Goliszewska, K.; Gryko, D. Vitamin B_{12} catalysed reactions. *Chem. Soc. Rev.* **2015**, *44*, 3391–3404. [CrossRef]
56. López-Rubio, J.J.; Padmanabhan, S.; Lázaro, J.M.; Salas, M.; Murillo, F.J.; Elías-Arnanz, M. Operator design and mechanism for CarA repressor-mediated down-regulation of the photoinducible *carB* operon in *Myxococcus xanthus*. *J. Biol. Chem.* **2004**, *279*, 28945–28953. [CrossRef] [PubMed]
57. López-Rubio, J.J.; Elías-Arnanz, M.; Padmanabhan, S.; Murillo, F.J. A repressor-antirepressor pair links two loci controlling light-induced carotenogenesis in *Myxococcus xanthus*. *J. Biol. Chem.* **2002**, *277*, 7262–7270. [CrossRef]
58. Navarro-Avilés, G.; Jiménez, M.A.; Pérez-Marín, M.C.; González, C.; Rico, M.; Murillo, F.J.; Elías-Arnanz, M.; Padmanabhan, S. Structural basis for operator and antirepressor recognition by *Myxococcus xanthus* CarA repressor. *Mol. Microbiol.* **2007**, *63*, 980–994. [CrossRef] [PubMed]
59. Pérez-Marín, M.C.; López-Rubio, J.J.; Murillo, F.J.; Elías-Arnanz, M.; Padmanabhan, S. The N-terminus of *Myxococcus xanthus* CarA repressor is an autonomously folding domain that mediates physical and functional interactions with both operator DNA and antirepressor protein. *J. Biol. Chem.* **2004**, *279*, 33093–33103. [CrossRef] [PubMed]
60. Pérez-Marín, M.C.; Padmanabhan, S.; Polanco, M.C.; Murillo, F.J.; Elías-Arnanz, M. Vitamin B_{12} partners the CarH repressor to downregulate a photoinducible promoter in *Myxococcus xanthus*. *Mol. Microbiol.* **2008**, *67*, 804–819. [CrossRef]
61. Jost, M.; Simpson, J.H.; Drennan, C.L. The transcription factor CarH safeguards use of adenosylcobalamin as a light sensor by altering the photolysis products. *Biochemistry* **2015**, *54*, 3231–3234. [CrossRef]
62. Kutta, R.J.; Hardman, S.J.; Johannissen, L.O.; Bellina, B.; Messiha, H.L.; Ortiz-Guerrero, J.M.; Elías-Arnanz, M.; Padmanabhan, S.; Barran, P.; Scrutton, N.S.; et al. The photochemical mechanism of a B_{12}-dependent photoreceptor protein. *Nat. Commun.* **2015**, *6*, 7907. [CrossRef]
63. Miller, N.A.; Kaneshiro, A.K.; Konar, A.; Alonso-Mori, R.; Britz, A.; Deb, A.; Glownia, J.M.; Koralek, J.D.; Mallik, L.; Meadows, J.H.; et al. The photoactive excited state of the B_{12}-based photoreceptor CarH. *J. Phys. Chem. B* **2020**, *124*, 10732–10738. [CrossRef]
64. Fernández-Zapata, J.; Pérez-Castaño, R.; Aranda, J.; Colizzi, F.; Polanco, M.C.; Orozco, M.; Padmanabhan, S.; Elías-Arnanz, M. Plasticity in oligomerization, operator architecture, and DNA binding in the mode of action of a bacterial B_{12}-based photoreceptor. *J. Biol. Chem.* **2018**, *293*, 17888–17905. [CrossRef]
65. Takano, H.; Mise, K.; Hagiwara, K.; Hirata, N.; Watanabe, S.; Toriyabe, M.; Shiratori-Takano, H.; Ueda, K. Role and function of LitR, an adenosyl B_{12}-bound light-sensitive regulator of *Bacillus megaterium* QM B1551, in regulation of carotenoid production. *J. Bacteriol.* **2015**, *197*, 2301–2315. [CrossRef]
66. Díez, A.I.; Ortiz-Guerrero, J.M.; Ortega, A.; Elías-Arnanz, M.; Padmanabhan, S.; de la García Torre, J. Analytical ultracentrifugation studies of oligomerization and DNA-binding of TtCarH, a *Thermus thermophilus* coenzyme B_{12}-based photosensory regulator. *Eur. Biophys. J.* **2013**, *42*, 463–476. [CrossRef]
67. Toda, M.J.; Mamun, A.A.; Lodowski, P.; Kozlowski, P.M. Why is CarH photolytically active in comparison to other B(12)-dependent enzymes? *J. Photochem. Photobiol. B Biol.* **2020**, *209*, 111919. [CrossRef]

68. Schwartz, P.A.; Frey, P.A. 5′-Peroxyadenosine and 5′-peroxyadenosylcobalamin as intermediates in the aerobic photolysis of adenosylcobalamin. *Biochemistry* **2007**, *46*, 7284–7292. [CrossRef] [PubMed]
69. Dowling, D.P.; Croft, A.K.; Drennan, C.L. Radical use of Rossmann and TIM barrel architectures for controlling coenzyme B_{12} chemistry. *Annu. Rev. Biophys.* **2012**, *41*, 403–427. [CrossRef] [PubMed]
70. Fontes, M.; Galbis-Martínez, L.; Murillo, F.J. A novel regulatory gene for light-induced carotenoid synthesis in the bacterium *Myxococcus xanthus*. *Mol. Microbiol.* **2003**, *47*, 561–571. [CrossRef] [PubMed]
71. Galbis-Martínez, L.; Galbis-Martínez, M.; Murillo, F.J.; Fontes, M. An anti-antisigma factor in the response of the bacterium *Myxococcus xanthus* to blue light. *Microbiology* **2008**, *154*, 895–904. [CrossRef] [PubMed]
72. Thomson, T.M.; Lozano, J.J.; Loukili, N.; Carrio, R.; Serras, F.; Cormand, B.; Valeri, M.; Diaz, V.M.; Abril, J.; Burset, M.; et al. Fusion of the human gene for the polyubiquitination coeffector UEV1 with *Kua*, a newly identified gene. *Genome Res.* **2000**, *10*, 1743–1756. [CrossRef] [PubMed]
73. Gao, J.; Ajjawi, I.; Manoli, A.; Sawin, A.; Xu, C.; Froehlich, J.E.; Last, R.L.; Benning, C. FATTY ACID DESATURASE4 of *Arabidopsis* encodes a protein distinct from characterized fatty acid desaturases. *Plant J.* **2009**, *60*, 832–839. [CrossRef]
74. He, A.; Dean, J.M.; Lodhi, I.J. Peroxisomes as cellular adaptors to metabolic and environmental stress. *Trends Cell Biol.* **2021**. [CrossRef]
75. Paul, S.; Lancaster, G.I.; Meikle, P.J. Plasmalogens: A potential therapeutic target for neurodegenerative and cardiometabolic disease. *Prog. Lipid Res.* **2019**, *74*, 186–195. [CrossRef] [PubMed]
76. Braverman, N.E.; Moser, A.B. Functions of plasmalogen lipids in health and disease. *Biochim. Biophys. Acta* **2012**, *1822*, 1442–1452. [CrossRef]
77. Ebenezer, D.L.; Fu, P.; Ramchandran, R.; Ha, A.W.; Putherickal, V.; Sudhadevi, T.; Harijith, A.; Schumacher, F.; Kleuser, B.; Natarajan, V. S1P and plasmalogen derived fatty aldehydes in cellular signaling and functions. *Biochim. Biophys. Acta Mol. Cell Biol. Lipids* **2020**, *1865*, 158681. [CrossRef]
78. Morand, O.H.; Zoeller, R.A.; Raetz, C.R. Disappearance of plasmalogens from membranes of animal cells subjected to photosensitized oxidation. *J. Biol. Chem.* **1988**, *263*, 11597–11606. [CrossRef]
79. Stadelmann-Ingrand, S.; Favreliere, S.; Fauconneau, B.; Mauco, G.; Tallineau, C. Plasmalogen degradation by oxidative stress: Production and disappearance of specific fatty aldehydes and fatty alpha-hydroxyaldehydes. *Free Radic. Biol. Med.* **2001**, *31*, 1263–1271. [CrossRef]
80. Dorninger, F.; Forss-Petter, S.; Wimmer, I.; Berger, J. Plasmalogens, platelet-activating factor and beyond—Ether lipids in signaling and neurodegeneration. *Neurobiol. Dis.* **2020**, *145*, 105061. [CrossRef]
81. Martínez-Argudo, I.; Ruiz-Vázquez, R.M.; Murillo, F.J. The structure of an ECF-sigma-dependent, light-inducible promoter from the bacterium *Myxococcus xanthus*. *Mol. Microbiol.* **1998**, *30*, 883–893. [CrossRef] [PubMed]
82. Whitworth, D.E.; Hodgson, D.A. Light-induced carotenogenesis in *Myxococcus xanthus*: Evidence that CarS acts as an anti-repressor of CarA. *Mol. Microbiol.* **2001**, *42*, 809–819. [CrossRef]
83. Lonetto, M.A.; Brown, K.L.; Rudd, K.E.; Buttner, M.J. Analysis of the *Streptomyces coelicolor sigE* gene reveals the existence of a subfamily of eubacterial RNA polymerase sigma factors involved in the regulation of extracytoplasmic functions. *Proc. Natl. Acad. Sci. USA* **1994**, *91*, 7573–7577. [CrossRef]
84. Lonetto, M.A.; Donohue, T.J.; Gross, C.A.; Buttner, M.J. Discovery of the extracytoplasmic function σ factors. *Mol. Microbiol.* **2019**, *112*, 348–355. [CrossRef]
85. Staron, A.; Sofia, H.J.; Dietrich, S.; Ulrich, L.E.; Liesegang, H.; Mascher, T. The third pillar of bacterial signal transduction: Classification of the extracytoplasmic function (ECF) sigma factor protein family. *Mol. Microbiol.* **2009**, *74*, 557–581. [CrossRef]
86. Browning, D.F.; Whitworth, D.E.; Hodgson, D.A. Light-induced carotenogenesis in *Myxococcus xanthus*: Functional characterization of the ECF sigma factor CarQ and antisigma factor CarR. *Mol. Microbiol.* **2003**, *48*, 237–251. [CrossRef]
87. Abellón-Ruiz, J.; Bernal-Bernal, D.; Abellán, M.; Fontes, M.; Padmanabhan, S.; Murillo, F.J.; Elías-Arnanz, M. The CarD/CarG regulatory complex is required for the action of several members of the large set of *Myxococcus xanthus* extracytoplasmic function sigma factors. *Environ. Microbiol.* **2014**, *16*, 2475–2490. [CrossRef] [PubMed]
88. Kohler, C.; Lourenco, R.F.; Avelar, G.M.; Gomes, S.L. Extracytoplasmic function (ECF) sigma factor σ^F is involved in *Caulobacter crescentus* response to heavy metal stress. *BMC Microbiol.* **2012**, *12*, 210. [CrossRef] [PubMed]
89. Masloboeva, N.; Reutimann, L.; Stiefel, P.; Follador, R.; Leimer, N.; Hennecke, H.; Mesa, S.; Fischer, H.M. Reactive oxygen species-inducible ECF σ factors of *Bradyrhizobium japonicum*. *PLoS ONE* **2012**, *7*, e43421. [CrossRef]
90. Stockwell, S.B.; Reutimann, L.; Guerinot, M.L. A role for *Bradyrhizobium japonicum* ECF16 sigma factor EcfS in the formation of a functional symbiosis with soybean. *Mol. Plant Microbe Interact.* **2012**, *25*, 119–128. [CrossRef]
91. Moraleda-Muñoz, A.; Pérez, J.; Fontes, M.; Murillo, F.J.; Muñoz-Dorado, J. Copper induction of carotenoid synthesis in the bacterium *Myxococcus xanthus*. *Mol. Microbiol.* **2005**, *56*, 1159–1168. [CrossRef]
92. Nicolás, F.J.; Ruiz-Vázquez, R.M.; Murillo, F.J. A genetic link between light response and multicellular development in the bacterium *Myxococcus xanthus*. *Genes Dev.* **1994**, *8*, 2375–2387. [CrossRef]
93. Moreno, A.J.; Fontes, M.; Murillo, F.J. *ihfA* gene of the bacterium *Myxococcus xanthus* and its role in activation of carotenoid genes by blue light. *J. Bacteriol.* **2001**, *183*, 557–569. [CrossRef] [PubMed]
94. Dame, R.T.; Rashid, F.M.; Grainger, D.C. Chromosome organization in bacteria: Mechanistic insights into genome structure and function. *Nat. Rev. Genet.* **2020**, *21*, 227–242. [CrossRef] [PubMed]

95. Dey, D.; Nagaraja, V.; Ramakumar, S. Structural and evolutionary analyses reveal determinants of DNA binding specificities of nucleoid-associated proteins HU and IHF. *Mol. Phylogenet. Evol.* **2017**, *107*, 356–366. [CrossRef] [PubMed]
96. Peñalver-Mellado, M.; García-Heras, F.; Padmanabhan, S.; García-Moreno, D.; Murillo, F.J.; Elías-Arnanz, M. Recruitment of a novel zinc-bound transcriptional factor by a bacterial HMGA-type protein is required for regulating multiple processes in *Myxococcus xanthus*. *Mol. Microbiol.* **2006**, *61*, 910–926. [CrossRef]
97. Elías-Arnanz, M.; Padmanabhan, S.; Murillo, F.J. The regulatory action of the myxobacterial CarD/CarG complex: A bacterial enhanceosome? *FEMS Microbiol. Rev.* **2010**, *34*, 764–778. [CrossRef]
98. García-Heras, F.; Padmanabhan, S.; Murillo, F.J.; Elías-Arnanz, M. Functional equivalence of HMGA- and histone H1-like domains in a bacterial transcriptional factor. *Proc. Natl. Acad. Sci. USA* **2009**, *106*, 13546–13551. [CrossRef] [PubMed]
99. Marino-Puertas, L.; Goulas, T.; Gomis-Rüth, F.X. Matrix metalloproteinases outside vertebrates. *Biochim. Biophys. Acta. Mol. Cell Res.* **2017**, *1864*, 2026–2035. [CrossRef]
100. Haugen, S.P.; Ross, W.; Gourse, R.L. Advances in bacterial promoter recognition and its control by factors that do not bind DNA. *Nat. Rev. Microbiol.* **2008**, *6*, 507–519. [CrossRef]
101. Chen, J.; Boyaci, H.; Campbell, E.A. Diverse and unified mechanisms of transcription initiation in bacteria. *Nat. Rev. Microbiol.* **2021**, *19*, 95–109. [CrossRef]
102. Nicolás, F.J.; Cayuela, M.L.; Martínez-Argudo, I.M.; Ruiz-Vázquez, R.M.; Murillo, F.J. High mobility group I(Y)-like DNA-binding domains on a bacterial transcription factor. *Proc. Natl. Acad. Sci. USA* **1996**, *93*, 6881–6885. [CrossRef]
103. Padmanabhan, S.; Elías-Arnanz, M.; Carpio, E.; Aparicio, P.; Murillo, F.J. Domain architecture of a high mobility group A-type bacterial transcriptional factor. *J. Biol. Chem.* **2001**, *276*, 41566–41575. [CrossRef]
104. Reeves, R. Nuclear functions of the HMG proteins. *Biochim. Biophys. Acta* **2010**, *1799*, 3–14. [CrossRef]
105. García-Heras, F.; Abellón-Ruiz, J.; Murillo, F.J.; Padmanabhan, S.; Elías-Arnanz, M. High-mobility-group a-like CarD binds to a DNA site optimized for affinity and position and to RNA polymerase to regulate a light-inducible promoter in *Myxococcus xanthus*. *J. Bacteriol.* **2013**, *195*, 378–388. [CrossRef]
106. Whitworth, D.E.; Bryan, S.J.; Berry, A.E.; McGowan, S.J.; Hodgson, D.A. Genetic dissection of the light-inducible *carQRS* promoter region of *Myxococcus xanthus*. *J. Bacteriol.* **2004**, *186*, 7836–7846. [CrossRef] [PubMed]
107. Bernal-Bernal, D.; Gallego-García, A.; García-Martínez, G.; García-Heras, F.; Jiménez, M.A.; Padmanabhan, S.; Elías-Arnanz, M. Structure-function dissection of *Myxococcus xanthus* CarD N-Terminal domain, a defining member of the CarD_CdnL_TRCF family of RNA polymerase interacting proteins. *PLoS ONE* **2015**, *10*, e0121322. [CrossRef] [PubMed]
108. Cayuela, M.L.; Elías-Arnanz, M.; Peñalver-Mellado, M.; Padmanabhan, S.; Murillo, F.J. The *Stigmatella aurantiaca* homolog of *Myxococcus xanthus* high-mobility-group A-type transcription factor CarD: Insights into the functional modules of CarD and their distribution in bacteria. *J. Bacteriol.* **2003**, *185*, 3527–3537. [CrossRef] [PubMed]
109. Selby, C.P.; Sancar, A. Structure and function of transcription-repair coupling factor. I. Structural domains and binding properties. *J. Biol. Chem.* **1995**, *270*, 4882–4889. [CrossRef]
110. García-Moreno, D.; Abellón-Ruiz, J.; García-Heras, F.; Murillo, F.J.; Padmanabhan, S.; Elías-Arnanz, M. CdnL, a member of the large CarD-like family of bacterial proteins, is vital for *Myxococcus xanthus* and differs functionally from the global transcriptional regulator CarD. *Nucleic Acids Res.* **2010**, *38*, 4586–4598. [CrossRef] [PubMed]
111. Stallings, C.L.; Stephanou, N.C.; Chu, L.; Hochschild, A.; Nickels, B.E.; Glickman, M.S. CarD is an essential regulator of rRNA transcription required for *Mycobacterium tuberculosis* persistence. *Cell* **2009**, *138*, 146–159. [CrossRef]
112. Iniesta, A.A.; García-Heras, F.; Abellón-Ruiz, J.; Gallego-García, A.; Elías-Arnanz, M. Two systems for conditional gene expression in *Myxococcus xanthus* inducible by isopropyl-β-D-thiogalactopyranoside or vanillate. *J. Bacteriol.* **2012**, *194*, 5875–5885. [CrossRef]
113. Gallego-García, A.; Mirassou, Y.; García-Moreno, D.; Elías-Arnanz, M.; Jiménez, M.A.; Padmanabhan, S. Structural insights into RNA polymerase recognition and essential function of *Myxococcus xanthus* CdnL. *PLoS ONE* **2014**, *9*, e108946. [CrossRef] [PubMed]
114. Yang, X.F.; Goldberg, M.S.; He, M.; Xu, H.; Blevins, J.S.; Norgard, M.V. Differential expression of a putative CarD-like transcriptional regulator, LtpA, in *Borrelia burgdorferi*. *Infect. Immun.* **2008**, *76*, 4439–4444. [CrossRef] [PubMed]
115. Gallego-García, A.; Iniesta, A.A.; Gonzalez, D.; Collier, J.; Padmanabhan, S.; Elías-Arnanz, M. *Caulobacter crescentus* CdnL is a non-essential RNA polymerase-binding protein whose depletion impairs normal growth and rRNA transcription. *Sci. Rep.* **2017**, *7*, 43240. [CrossRef]
116. Henry, K.K.; Ross, W.; Myers, K.S.; Lemmer, K.C.; Vera, J.M.; Landick, R.; Donohue, T.J.; Gourse, R.L. A majority of Rhodobacter sphaeroides promoters lack a crucial RNA polymerase recognition feature, enabling coordinated transcription activation. *Proc. Natl. Acad. Sci. USA* **2020**, *117*, 29658–29668. [CrossRef]
117. Boyaci, H.; Chen, J.; Jansen, R.; Darst, S.A.; Campbell, E.A. Structures of an RNA polymerase promoter melting intermediate elucidate DNA unwinding. *Nature* **2019**, *565*, 382–385. [CrossRef] [PubMed]
118. Woldemeskel, S.A.; Daitch, A.K.; Alvarez, L.; Panis, G.; Zeinert, R.; Gonzalez, D.; Smith, E.; Collier, J.; Chien, P.; Cava, F.; et al. The conserved transcriptional regulator CdnL is required for metabolic homeostasis and morphogenesis in *Caulobacter*. *PLoS Genet.* **2020**, *16*, e1008591. [CrossRef] [PubMed]
119. Galbis-Martínez, M.; Fontes, M.; Murillo, F.J. The high-mobility group A-type protein CarD of the bacterium *Myxococcus xanthus* as a transcription factor for several distinct vegetative genes. *Genetics* **2004**, *167*, 1585–1595. [CrossRef]

120. Bernal-Bernal, D.; Abellón-Ruiz, J.; Iniesta, A.A.; Pajares-Martínez, E.; Bastida-Martínez, E.; Fontes, M.; Padmanabhan, S.; Elías-Arnanz, M. Multifactorial control of the expression of a CRISPR-Cas system by an extracytoplasmic function sigma/anti-sigma pair and a global regulatory complex. *Nucleic Acids Res.* **2018**, *46*, 6726–6745. [CrossRef] [PubMed]
121. León, E.; Navarro-Avilés, G.; Santiveri, C.M.; Flores-Flores, C.; Rico, M.; González, C.; Murillo, F.J.; Elías-Arnanz, M.; Jiménez, M.A.; Padmanabhan, S. A bacterial antirepressor with SH3 domain topology mimics operator DNA in sequestering the repressor DNA recognition helix. *Nucleic Acids Res.* **2010**, *38*, 5226–5241. [CrossRef] [PubMed]
122. López-García, P.; Moreira, D. The Syntrophy hypothesis for the origin of eukaryotes revisited. *Nat. Microbiol.* **2020**, *5*, 655–667. [CrossRef]
123. García-Moreno, D.; Polanco, M.C.; Navarro-Avilés, G.; Murillo, F.J.; Padmanabhan, S.; Elías-Arnanz, M. A vitamin B_{12}-based system for conditional expression reveals *dksA* to be an essential gene in *Myxococcus xanthus*. *J. Bacteriol.* **2009**, *191*, 3108–3119. [CrossRef]
124. Chatelle, C.; Ochoa-Fernandez, R.; Engesser, R.; Schneider, N.; Beyer, H.M.; Jones, A.R.; Timmer, J.; Zurbriggen, M.D.; Weber, W. A green-light-responsive system for the control of transgene expression in mammalian and plant cells. *ACS Synth. Biol.* **2018**, *7*, 1349–1358. [CrossRef] [PubMed]
125. Kainrath, S.; Stadler, M.; Reichhart, E.; Distel, M.; Janovjak, H. Green-light-induced inactivation of receptor signaling using cobalamin-binding domains. *Angew. Chem. Int. Ed. Engl.* **2017**, *56*, 4608–4611. [CrossRef] [PubMed]
126. Jiang, B.; Liu, X.; Yang, C.; Yang, Z.; Luo, J.; Kou, S.; Liu, K.; Sun, F. Injectable, photoresponsive hydrogels for delivering neuroprotective proteins enabled by metal-directed protein assembly. *Sci. Adv.* **2020**, *6*, eabc4824. [CrossRef]
127. Wang, R.; Yang, Z.; Luo, J.; Hsing, I.M.; Sun, F. B_{12}-dependent photoresponsive protein hydrogels for controlled stem cell/protein release. *Proc. Natl. Acad. Sci. USA* **2017**, *114*, 5912–5917. [CrossRef] [PubMed]
128. Xu, D.; Ricken, J.; Wegner, S. Turning cell adhesions ON or OFF with high spatiotemporal precision using the green light responsive protein CarH. *Chemistry* **2020**. [CrossRef]
129. Jain, I.H.; Calvo, S.E.; Markhard, A.L.; Skinner, O.S.; To, T.L.; Ast, T.; Mootha, V.K. Genetic screen for cell fitness in high or low oxygen highlights mitochondrial and lipid metabolism. *Cell* **2020**, *181*, 716–727 e711. [CrossRef]
130. Zou, Y.; Henry, W.S.; Ricq, E.L.; Graham, E.T.; Phadnis, V.V.; Maretich, P.; Paradkar, S.; Boehnke, N.; Deik, A.A.; Reinhardt, F.; et al. Plasticity of ether lipids promotes ferroptosis susceptibility and evasion. *Nature* **2020**, *585*, 603–608. [CrossRef]
131. Cui, W.; Liu, D.; Gu, W.; Chu, B. Peroxisome-driven ether-linked phospholipids biosynthesis is essential for ferroptosis. *Cell Death Differ.* **2021**. [CrossRef]

microorganisms

MDPI

Article

Predatory Bacteria Select for Sustained Prey Diversity

Ramith R. Nair [1,2,*] and Gregory J. Velicer [2]

1 Department of Medical Biochemistry and Microbiology, Uppsala University, 75234 Uppsala, Sweden
2 Institute for Integrative Biology, ETH Zürich, 8092 Zürich, Switzerland; gregory.velicer@env.ethz.ch
* Correspondence: ramith_nair@hotmail.com

Abstract: Predator impacts on prey diversity are often studied among higher organisms over short periods, but microbial predator-prey systems allow examination of prey-diversity dynamics over evolutionary timescales. We previously showed that *Escherichia coli* commonly evolved minority mucoid phenotypes in response to predation by the bacterial predator *Myxococcus xanthus* by one time point of a coevolution experiment now named MyxoEE-6. Here we examine mucoid frequencies across several MyxoEE-6 timepoints to discriminate between the hypotheses that mucoids were increasing to fixation, stabilizing around equilibrium frequencies, or heading to loss toward the end of MyxoEE-6. In four focal coevolved prey populations, mucoids rose rapidly early in the experiment and then fluctuated within detectable minority frequency ranges through the end of MyxoEE-6, generating frequency dynamics suggestive of negative frequency-dependent selection. However, a competition experiment between mucoid and non-mucoid clones found a predation-specific advantage of the mucoid clone that was insensitive to frequency over the examined range, leaving the mechanism that maintains minority mucoidy unresolved. The advantage of mucoidy under predation was found to be associated with reduced population size after growth (productivity) in the absence of predators, suggesting a tradeoff between productivity and resistance to predation that we hypothesize may reverse mucoid vs non-mucoid fitness ranks within each MyxoEE-6 cycle. We also found that mucoidy was associated with diverse colony phenotypes and diverse candidate mutations primarily localized in the exopolysaccharide operon *yjbEFGH*. Collectively, our results show that selection from predatory bacteria can generate apparently stable sympatric phenotypic polymorphisms within coevolving prey populations and also allopatric diversity across populations by selecting for diverse mutations and colony phenotypes associated with mucoidy. More broadly, our results suggest that myxobacterial predation increases long-term diversity within natural microbial communities.

Keywords: pretator-prey coevolution; antagonism; mucoidy; predatory bacteria; bacterial predation; prey diversity; negative frequency dependence; experimental evolution; MyxoEE-6

Citation: Nair, R.R.; Velicer, G.J. Predatory Bacteria Select for Sustained Prey Diversity. *Microorganisms* **2021**, *9*, 2079. https://doi.org/10.3390/microorganisms9102079

Academic Editor: David Whitworth

Received: 17 July 2021
Accepted: 23 September 2021
Published: 2 October 2021

1. Introduction

Predation is one of the most common forms of inter-specific antagonism [1]. Under predation pressure, prey face the dual challenges of optimizing their own acquisition and use of resources for growth and reproduction while avoiding being killed or injured by predators. This dilemma has been shown to play an important role in the ecology and evolution of diverse prey species, including among plants [2], animals [3] and microorganisms [4].

Bacteria fall prey to a wide variety of predators, including unicellular eukaryotes [5], amoebae [6], nematodes [7], and even other bacteria [8,9]. Several studies have shown that over short time periods, microbial predators can elicit phenotypic responses providing resistance against predatory killing. These include filamentation [10], biofilm formation [11], sporulation [12] and production of various extracellular compounds [13] (reviewed in detail in [14]). Although these mechanisms can help thwart predation, many are part of a repertoire of responses that also protect against other stresses, making the degree to which they are selected specifically by predation unclear. Longer studies of predator-prey interactions make it possible to identify evolutionary-scale responses of prey specific to interaction with predators and to characterize their temporal evolutionary dynamics.

Long-term coevolution studies have examined the emergence and subsequent evolution of adaptive defensive traits of bacteria under attack by phage [15–17], but few have done so with predatory bacteria. Such coevolutionary studies with phage have shown that bacteria adapt to phage predation by modifying molecules involved in phage adsorption or by producing extracellular polysaccharides to restrict access to the cell surface. However, the reproductive rate of phage is generally much higher than that of its prey, such that dynamics and mechanisms of coevolution between bacteria and phage likely differ greatly from those between bacterial prey and slower-growing predatory bacteria.

Myxobacteria are soil- and sediment-dwelling bacteria that kill and consume diverse other microbes, including both Gram+ and Gram– bacteria and fungi, by mechanisms that remain poorly understood [18–20]. Because of this broad prey range, predation by myxobacteria is predicted to play significant roles in shaping the composition, structure and evolution of complex microbial communities [21].

In a predator-prey coevolution experiment with *Escherichia coli* as prey and the myxobacterium *Myxococcus xanthus* as predator recently named MyxoEE-6 [22], we previously showed that coevolving prey were under selection both for parallel losses of function in a prey outer-membrane protein (OmpT) and favoring the increase of genotypes that generate a mucoid-colony phenotype [23]. Mucoid colonies are characterized by increased opacity, generally lighter pigmentation and convex colony surfaces relative to non-mucoid colonies, although detailed mucoid-colony phenotypes may vary among conspecifics. Mucoidy is achieved by increased secretion of extracellular polysaccharides and can evolve in response to a number of biotic and abiotic challenges, including bacteriophages [15], antibiotics [24], macrophages [25] and menthol [26]. Mucoid colony-forming *E. coli* cells were found to arise frequently in MyxoEE-6 populations co-evolving with *M. xanthus*, but were largely absent from prey-only control populations [23]. Additionally, mucoidy was associated with reduced swarming and killing by the predator.

Our first study of MyxoeEE-6 and many earlier studies have shown that predation often strongly impacts prey diversity [23,27–30], but longer-term fates and dynamics of such predation-induced diversity are underexplored. Here we test whether mucoid lineages in MyxoEE-6 were on frequency trajectories predictive of long-term fixation of the phenotype or rather reveal the operation of evolutionary mechanisms preventing fixation, for example negatively frequency-dependent selection or clonal interference [31–33]. We then test for negative frequency dependence (NFD) of fitness in one clone pair and for cost of mucoidy that might promote extended maintenance of mucoid/non-mucoid polymorphisms. Finally, we characterize inter-population genetic and phenotypic diversity mediated by predation.

2. Materials and Methods

2.1. Predator-Prey Coevolution

MyxoEE-6 was conducted as described in Nair et al. (2019) [23]. Briefly, replicate populations of *M. xanthus* strain DK3470 and *E. coli* strain MG1655 (along with prey-only and predator-only controls) were paired and spread with glass beads on 8 mL prey-growth agar (1$\times$ M9 salts, 2 mM $MgSO_4$, 0.1 mM $CaCl_2$, 0.2% glucose, 1.5% agar) in 50 mL conical flasks. The flasks were incubated at 32 °C for 84 h, after which they were harvested by adding 5 mL TPM buffer and shaking on an orbital shaker set at 300 rpm for 15 min. 1% of the surviving community was transferred to fresh minimal media and the cycle was repeated 25 times, allowing $\sim$166 generations of growth.

2.2. Mucoid Frequency Estimation and Clone Isolation

Mucoid frequency estimation. Evolved *E. coli* colonies that were of lighter pigmentation, more opaque and more convex relative to the ancestral colony phenotype on LB agar were identified as mucoid. To determine the frequency of mucoids in each replicate MyxoEE-6 community/population at different time points, aliquots of frozen stocks from respective lineages and time-points were thawed in 100 µL TPM buffer, diluted and plated onto LB agar plates. Following overnight incubation at 32 °C, 90% humidity, mucoid and non-mucoid colonies were distinguished and counted. Cycle 18 and cycle 25 populations were sampled over successive sampling periods, with each replicate sampling for each cycle performed independently at different times. Samples from cycles 6, 10, 14, 16, 20 and 22 of *E. coli* populations ME4, ME8, ME11 and ME12 were collected and plated together in the same sets of at least 3 temporally separated biological replicates. When zero mucoid colonies were present on the plate selected for counting from a dilution series, we estimated a hypothetical maximum mucoid frequency as [1/(colony N + 1)] to mimic a scenario in which the actual colony count N for the plate was increased by one and the additional hypothetical colony was mucoid. The four populations with the highest mucoid frequencies at cycle 18 (ME4, ME8, ME11 and ME12) were selected for examination at more time points to allow the greatest opportunity to accurately resolve frequency dynamics. The colonies scored to assess mucoid frequencies were not subsequently cultured or stored frozen.

Isolation of cycle 18 mucoid and non-mucoid clones. One mucoid and one non-mucoid clone each were isolated from cycle-18 co-evolved populations in which mucoids were detected and subsequently stored frozen. The clone pair from ME4 was used for the direct competition experiment measuring mucoid vs non-mucoid relative fitness and all nine clone pairs were used in the survival-under-predation and productivity assays and for colony-phenotype imaging. Single colonies of each phenotype were picked and streaked onto fresh LB agar plates to isolate a sub-colony and thereby purge any genetic variation that may have been present in the colony from the original plating. Isolated colonies of each phenotype were picked, grown to high density in LB liquid and samples of the resulting cultures were stored as frozen stocks (20% glycerol) for further experiments. The same mucoid and non-mucoid clone from each co-evolved population were used in the productivity and predation assays described below and were used for imaging colony phenotypes. For colony-phenotype imaging, aliquots of frozen stocks from mucoid and non-mucoid clones from nine different lineages were thawed in 100 µL TPM buffer and 10 µL of the suspension was spotted on an LB agar plate. The resulting colony was imaged after overnight growth at 32 °C using Olympus SZX16 stereomicroscope at 0.8X magnification. The same mucoid and non-mucoid clones from cycle 18 population ME4 were also used in the relative fitness assay.

2.3. Productivity and Predation Assays

Experiments measuring *E. coli* mucoid vs. non-mucoid relative fitness in the presence and absence of *M. xanthus*, survival in the presence of *M. xanthus* and population size after 25 h of incubation (productivity) were performed under the same abiotic conditions as MyxoEE-6 [23]. *E. coli* cells of each type were grown overnight in LB liquid at 32 °C, 300 rpm, then diluted to OD_{600} 1.0 with LB and subsequently diluted 1:100 with TPM liquid buffer immediately prior to initiating the relevant assay. Initial and final population sizes were determined by dilution plating onto LB agar and counting colonies after overnight incubation at 32 °C, 90% humidity. *M. xanthus* was grown in 8 mL CTT liquid (10 g/L casitone, 10 mM Tris pH 8.0, 8 mM $MgSO_4$, 1 mM KPO_4) in 50 mL flasks at 32 °C, 300 rpm until mid-exponential phase, when cultures were centrifuged (5000 rpm, 15 min) and resuspended in TPM buffer to a density of $\sim 10^9$ cells/mL. 50 µL of the diluted prey suspension were then either mixed with 50 µL of *M. xanthus* cell suspension at $\sim 10^9$cells/mL or 50 µL TPM buffer (productivity assays and competitions without predator) before being spread onto MyxoEE-6 prey-growth agar [23]. For the relative-fitness competition experiment, one mucoid and one non-mucoid clone each isolated at the end of MyxoEE-6 cycle 18 from the coevolution treatment ME4 were used. In those experiments, cultures of the paired *E. coli* competitors were mixed at the specified ratios after the adjustment to OD_{600} 1.0 prior to proceeding as described above. Cultures were harvested as during MyxoEE-6. The selection rate constant was calculated as in [34] using the following formula:

$$\text{sij} = \frac{1}{t}\left[ln\left(\frac{N_i(t)}{N_i(0)}\right) - ln\left(\frac{N_j(t)}{N_j(0)}\right)\right], \quad (1)$$

where $N_i(0)$ and $N_i(t)$ are initial and final population densities of the mucoid clone, while $N_j(0)$ and $N_j(t)$ are initial and final population densities of the competing non-mucoid clone respectively.

2.4. Genomic Data

Candidate mutations for causation of mucoidy were determined by comparing whole-genome sequences from two non-mucoid clones and one mucoid clone from four co-evolved populations after cycle 25 (ME4, ME8, ME11, ME12). The genomic data for these 12 clones have already been published [23] and made available elsewhere (deposited in the SRA database under BioProject accession PRJNA551936 (BioSample accessions SAMN12169214—SAMN12169315)).

2.5. Statistical Analysis

Frequency dependence of the selection rate constant for competition experiments between mucoid and non-mucoid clones from population ME4 was tested using one-way ANOVA with starting frequency (0.1, 0.25, 0.5) as a factor. We tested for differences in average productivity as well as killing resistance between mucoid vs. non-mucoid clones using Welch's two-sided, two-sample *t*-tests. We tested for differences in productivity and killing resistance among mucoid clones from different coevolving populations with one-way ANOVA with population ID as a factor in both cases. Statistical analyses were performed in R [35] using R studio (version 1.3.1093) and plotted with the package ggplot2 version 3.3.4 [36].

3. Results

3.1. Mucoid Phenotypes Rise to and Remain within Intermediate Frequency Ranges

In the first study of MyxoEE-6, we reported parallel emergence of mucoid colony-forming variants in *E. coli* prey among ten out of twelve replicate populations that had co-evolved with *M. xanthus* for 18 cycles [23], whereas mucoids were largely absent from control *E. coli* populations that evolved without predators. This pattern strongly suggested that most or all mucoid genotypes detected among the co-evolved populations rose to high frequency due to selection rather than neutral drift. Indeed, mucoidy was found to be associated with reduced susceptibility to predation. However, mucoid colonies were present only as a minority after cycle 18, leaving open the question of whether, toward the end of MyxoEE-6, mucoids were (i) sweeping to fixation, (ii) decreasing toward loss due to clonal interference from non-mucoid adaptive mutants, or (iii) being maintained long-term by balancing selection while fluctuating around intermediate equilibrium frequencies.

In this study, we first compared mucoid frequencies in all MyxoEE-6 *E. coli* populations, including the 12 replicate populations that co-evolved with *M. xanthus* (ME1–ME12) and the six control replicate populations that evolved in the absence of *M. xanthus* (E1–E6), between cycles 18 and the end of the MyxoEE-6 experiment at cycle 25. Although mucoids appear to have generally decreased in frequency in many co-evolved prey populations from cycle 18 to 25, they were nonetheless present above our levels of detection in nine of the twelve co-evolved populations at cycle 25 (Figure S1). Between cycles 18 and 25, mucoids dropped below the limit of detection in two populations (ME7 and ME9) and newly rose to a detectable level in one population in which no mucoids had been found at cycle 18 (ME2, Figure S1). Thus, in total across the two time points, mucoid colony variants were observed in eleven out of the twelve coevolving prey populations and were at frequencies above the limit of detection at both time points in most. In contrast, among the six prey-only control populations, mucoidy was detected in only one replicate sample of one population at one time point (Figure S1).

The cycle 25 results did not strongly suggest trajectories toward either fixation or loss of mucoids. We thus addressed the above hypotheses more rigorously for the four prey populations with the highest mucoid frequencies at cycle 18 by examining their mucoid frequencies at six additional MyxoEE-6 timepoints (Figure 1). In all four populations, mucoids increased several orders of magnitude from zero to detectable frequencies already by the end of cycle 6, indicating strong selection favoring mucoidy. After this commonality of rapid early increase, mucoids in the four populations fluctuated within an intermediate range of detectable minority frequencies (Figure 1).

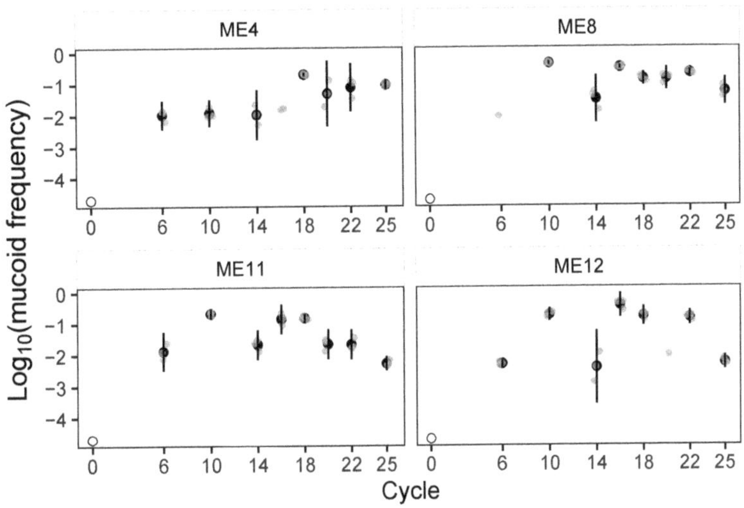

Figure 1. Mucoid prey rose to and persisted within intermediate frequency ranges in MyxoEE-6 in four focal populations. Frequencies of mucoid variants among the four focal coevolving prey populations estimated hypothetically at the start of MyxoEE-6 (cycle 0, open circles) and estimated directly by sampling at the end of eight MyxoEE-6 cycles (see Methods). A mucoid frequency of $\sim 10^{-5}$ at cycle 0 is shown for a hypothetical scenario in which there was one mucoid cell among the $\sim 10^{5}$ starting prey cells for each *E. coli* population. Grey dots show individual replicate estimates, black dots indicate cross-replicate means and error bars show 95% confidence intervals (*t*-distribution, three temporally separated biological replicates).

After cycle 6, however, each of the four focal populations showed a qualitatively unique pattern of subsequent mucoid-frequency dynamics. Mucoids in the ME4 community exhibited the most stable dynamics, remaining at frequencies near 0.01 from cycles 6–16, rising $\sim$10-fold by cycle 18 and remaining at frequencies near 0.1 for the duration of MyxoEE-6. Mucoids in ME8, ME11 and ME12 showed strikingly similar dynamics through cycle 16, all increasing to frequencies above 0.1 by cycle 10, subsequently decreasing by cycle 14 and increasing in parallel again by cycle 16 before diverging more in their detailed dynamics through the remaining cycles.

It is noteworthy that on the one hand, none of the four populations showed any indication of mucoids approaching fixation and, on the other hand, after cycle 25, mucoids in all populations remained within or very near the respective mucoid frequency range covered from cycles 6–22. Thus, despite late decreases in three populations (ME8, ME11 and ME12), the observed patterns are consistent with, and in our view suggestive of, extended maintenance of minority mucoids by NFD of mucoid fitness.

3.2. A Single-Cycle Competition Experiment Reveals a Predation-Specific Advantage to Mucoidy but Not NFD

NFD of fitness might be generated by direct interactions between mucoid and non-mucoid cells within a given MyxoEE-6 growth cycle. For example, costly production of exopolysaccharides by mucoids might confer some degree of social protection from predation and thus a fitness advantage to non-mucoids, with the degree of benefit conferred correlating with mucoid frequency. In this cheating scenario, NFD-mediated maintenance of the mucoid/non-mucoid polymorphism should be manifested by a reversal of fitness ranks (e.g., [37,38]) between mucoids and non-mucoids simply as a function of frequency within a single MyxoEE-6 growth cycle.

In our first study of MyxoEE-6, we reported that fewer cells of a mucoid clone isolated from the coevolving ME4 prey population at cycle 18 are killed by the ancestral predator compared to a contemporary non-mucoid isolate [23], but direct competition experiments between these prey isolates were not performed. Here we directly competed these same two isolates in the presence and absence of predators to both confirm a predicted direct relative fitness advantage of the mucoid isolate specific to predation pressure and to test whether this fitness advantage is frequency dependent within a single competition cycle due to social cheating.

The mucoid and non-mucoid prey clones were mixed at initial starting mucoid frequencies of 0.1, 0.25 and 0.5 and allowed to compete while growing under the same the experimental conditions as MyxoEE-6, except that the predator (a clone of *M. xanthus* also isolated from the ME4 community after cycle 18) was either absent or added at a population size of $\sim 5 \times 10^7$ predator cells to impose predation pressure. In these assays, the mixed *E. coli* populations generally increased more than 100-fold during the competition period. As expected, the mucoid genotype had higher fitness than the non-mucoid clone in the presence of *M. xanthus* but not in its absence (Figure 2, one-way ANOVA, $F_{1,16} = 235.1$, $p = 5.49 \times 10^{-11}$). However, NFD of fitness within a single competition cycle was not detected for this particular pair of competitors (one-way ANOVA, $F_{2,6} = 0.644$, $p = 0.558$). These results thus do not support the hypothesis that mucoids fail to reach fixation due to cheating-mediated NFD. However, we note that this outcome does not generally exclude the NFD hypothesis for other experimental conditions or for other mucoid and non-mucoid competitors.

3.3. Mucoidy Is Associated with Low Prey Productivity in the Absence of Predators

Adaptations of prey that decrease susceptibility to predation often come at a cost to other components of fitness. We previously found no cost of mucoidy to growth rate from 5–20 h after inoculation into the MyxoEE-6 selective regime for a pair of ME4 cycle 18 clones in the absence of predators [23]. However, final population-size productivity as well as rate of population increase can impact overall fitness and we hypothesized that mucoidy might come at cost to productivity. To explore this possibility, we examined one mucoid and one non-mucoid clone each from nine of the cycle 18 coevolving populations in which mucoids were detected (ME3, ME4, ME5, ME6, ME7, ME8, ME9, ME11, ME12).

We first confirmed that the mucoid clones are on average less susceptible to killing by *M. xanthus* than the corresponding non-mucoid clones from the same population (Figure 3, Welch's two-sample *t*-test: $t_{15.8} = 2.558$, $p = 0.011$), as was expected from previous results with the clone pair from population ME4 [23]. To examine the productivity-cost hypothesis, the clones were grown under MyxoEE-6 abiotic conditions for 25 h in the absence of predators and their final population sizes (productivity) determined. The average productivity of mucoid clones was found to be $\sim$27% lower than that of non-mucoid clones (Figure 4, Welch's two-sample *t*-test: $t_{14.18} = -3.5636$, $p = 0.0015$). Thus, mucoidy protects against predation, but at a cost to total productivity in the absence of predation. This productivity cost of mucoidy led us to speculate that it may mediate a form of NFD in which mucoid frequency impacts total *E. coli* productivity, which, along with and decreased susceptibility to predation by mucoids, impacts predator productivity at the end of a growth cycle, which in turn impacts mucoid vs non-mucoid relative fitness in the subsequent growth cycle. We elaborate this hypothesis further in the Discussion.

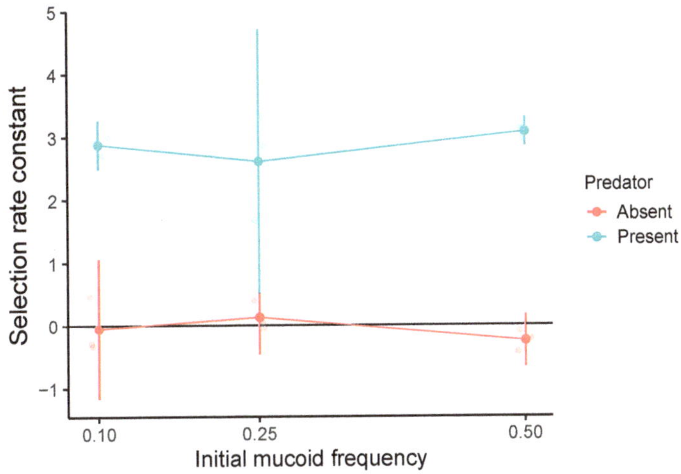

Figure 2. A mucoid clone outcompetes a non-mucoid clone only under predation pressure but does not exhibit negative frequency dependence. Estimates of the fitness of a mucoid clone isolated from ME4 at cycle 18 in competition relative to a contemporary non-mucoid clone (as represented by the selection-rate constant) across three initial frequencies are shown. x-axis values depict the estimated starting frequencies for the mucoid clone at the beginning of the experiment. Positive and negative values indicate estimates that the mucoid clone has higher or lower fitness than the non-mucoid clone, respectively. Smaller and larger dots represent individual-replicate and mean values, respectively. Error bars show 95% confidence intervals from three biological replicates (*t*-distribution).

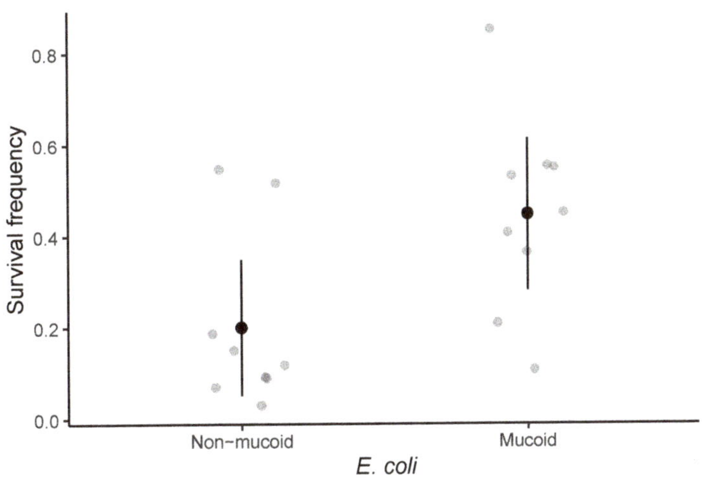

Figure 3. Mucoidy is associated with higher survival of encounters with *M. xanthus*. Frequencies of prey surviving encounters with *M. xanthus* are shown. Grey dots are mean survival frequency for clones from each of nine coevolving populations from which the clones were isolated at the end of cycle 18. Three independent estimates were generated for each population (Figure 5b). Black dots are means across the nine populations for each prey type and error bars show 95% confidence intervals (*t*-distribution).

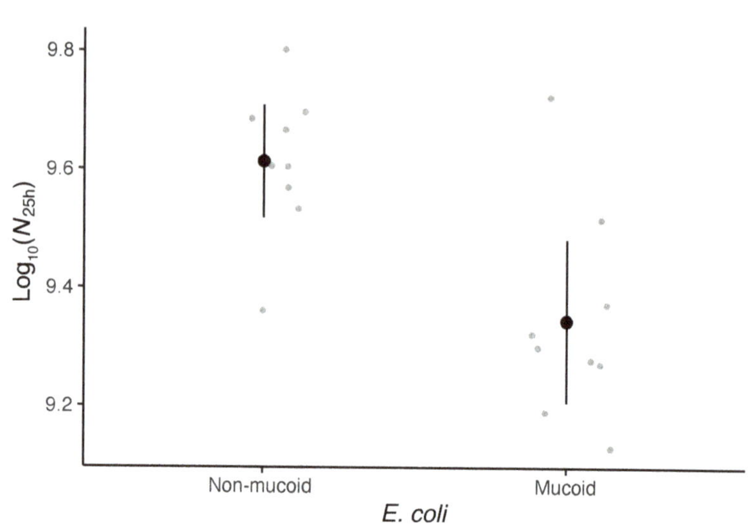

Figure 4. Mucoids have lower population productivity than non-mucoids in the absence of predators. Log-transformed (productivity) is shown. Grey dots are mean values from nine coevolving populations from which the clones were isolated at the end of cycle 18. Three independent estimates were generated for each population (Figure 5c). Black dots are means across the nine populations for each prey type and error bars show 95% confidence intervals from nine replicate populations (*t*-distribution).

3.4. Mucoidy Is Phenotypically Variable

We observed visually that colonies of the nine mucoid clones isolated from different lineages at the end of cycle 18 were phenotypically diverse, varying with regard to colony size, opacity, color and degree of phenotypic differentiation from non-mucoid colonies from their respective populations (Figure 5). We therefore tested for quantitative variation among these clones with respect to susceptibility to predation and productivity. While the mucoid clones are collectively less susceptible to killing by *M. xanthus* than non-mucoids (Figure 3), the degree of resistance to killing varied significantly among the nine mucoid clones tested here (Figure 5b, one-way ANOVA, $F_{8,18}$ = 3.988, p = 0.007). Additionally, while mucoids collectively have lower productivity than non-mucoids in the absence of predation (Figure 4), the isolated mucoid clones varied significantly in their productivity relative to contemporaneous non-mucoid clones isolated from the same prey population (Figure 5c), one-way ANOVA, $F_{8,18}$ = 4.4158, p = 0.004). Thus, while mucoidy in general is clearly selected by bacterial predation, the broader categorical phenotype can be associated with a diversity of detailed colony phenotypes and predation-related parameter values.

3.5. Candidate Mutations for Mucoidy

The observed variation in mucoid phenotypes suggested that distinct prey lineages may have followed different genomic routes to mucoidy, with distinct causal mutations having different phenotypic effects. Alternatively, the variable mucoid phenotypes might have resulted from differences in epistatic interactions between shared mutations or mutation targets (genes or gene pathways) and other evolved mutations in the same genetic background. To generate hypotheses regarding genetic causation of mucoidy in MyxoEE-6, we compared whole-genome sequences of one mucoid and two non-mucoid clones each from the four focal coevolved prey populations examined here for mucoidy dynamics across MyxoEE-6 (ME4, ME8, ME11, ME12) [23]. Mucoid clones from all four populations each had a mutation in one gene in the *yjbEFGH* operon, with only one gene (*yjbH*) mutated in more than one population (Table 1). In contrast, no mutations in this operon were found in any of the non-mucoid colonies, strongly suggesting that mutations in this operon confer mucoidy. This operon encodes proteins involved in the production of an uncharacterized extracellular polysaccharide [39]. Additionally, the mucoid clone from ME4 also has a mutation upstream of the gene *rcsA* (Table 1). Because *rcsA* is a transcriptional regulator of colanic acid capsular biosynthesis [40], this mutation may contribute to mucoidy in this clone.

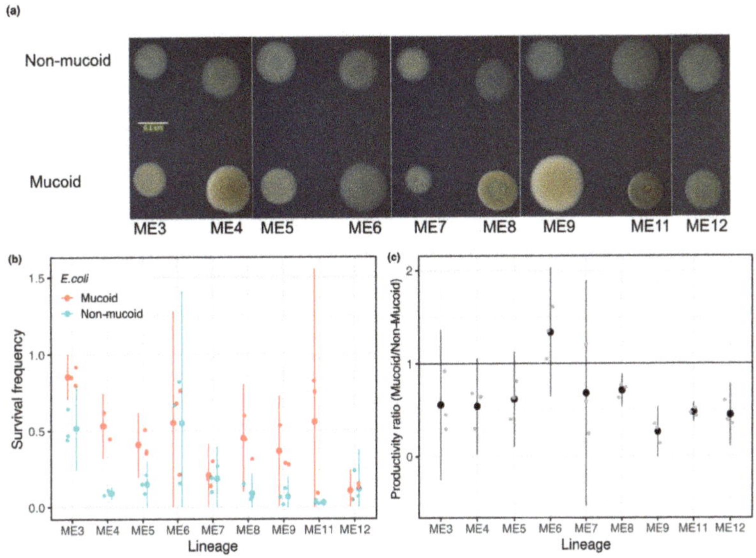

Figure 5. Allopatric variation among mucoid prey. (**a**) Observed mucoid (bottom row) and non-mucoid (top row) colony phenotypes on LB agar for the respective clones isolated from nine coevolving populations at the end of cycle 18 and used in the experiments reported in Figures 3, 4 and 5b,c. Phenotypic differences between the mucoid isolates and between mucoid vs. non-mucoid clones from the same population are yet more pronounced upon direct visual and microscopic observation than in the two-dimensional images shown here. (**b**) Frequencies of prey surviving encounters with *M. xanthus* are shown. (**c**) The relative productivity (population size after 25 hr) of single mucoid vs non-mucoid clones isolated from nine populations at the end of cycle 18 is shown. A value of 1 (solid black line) indicates equal productivity. (**b**,**c**) Small dots are values from individual replicates and large dots are inter-replicate means. Errors bars show 95% confidence intervals (t-distribution, n = 3).

Table 1. Mutation candidates for causation of mucoidy in cycle 25 mucoid isolates from the four focal co-evolved populations.

Gene	Mutation	Gene function	Population
rcsA	deletion (1 bp)	transcriptional regulator of colanic acid capsular biosynthesis	ME4
yjbG	SNP	extracellular polysaccharide export outer-membrane associated protein	ME4
yjbE	insertion (+9 bp)	extracellular polysaccharide production threonine-rich protein	ME8
yjbH	SNP	DUF940 family extracellular polysaccharide protein	ME11
yjbH	insertion (+8 bp)	DUF940 family extracellular polysaccharide protein	ME12

4. Discussion

By imposing predatory selection on prey, predators can increase diversity within prey communities and populations by diverse mechanisms. For example, keystone predators can promote diversity by targeting the most dominant prey species and thereby increasing resources available for other prey species [27–29,41,42]. Predation can ameliorate competitive exclusion among competing prey species by promoting prey that are less competitive at low predation pressure but less susceptible to predation by a dominant predator [10,43]. This can result in a negative feedback loop in which predation drives increase of less-susceptible prey, eventually resulting in reduced predator population size (and thus reduced predation pressure), which in turn reverses prey fitness ranks back in favor of prey that are more competitive at low predation pressure [28]. Predators can also maintain polymorphisms among prey through apostatic selection [30,44,45] and across environmental gradients by selecting for different prey alleles in different environmental contexts [46]. While such effects of predation on prey diversity have been investigated in a broad array of organisms, predatory bacteria are understudied in this regard.

Having previously shown that predatory bacteria select for mucoid variants of prey during coevolution [23], here we tested whether mucoid frequency dynamics indicate or suggest longer-term fixation, loss or intermediate persistence of mucoidy, especially in four focal populations of MyxoEE-6. Mucoid frequencies increased several orders of magnitude early in the experiment until reaching detectable frequencies already by cycle 6 (Figure 1). At this rate of increase, mucoids should have easily reached fixation during MyxoEE-6 in the absence of clonal interference or negative frequency dependence, but did not. Among the four focal populations, mucoids ceased increasing after reaching detectable frequencies and never reached majority status at sampled time points. After reaching their maximum frequencies, rather than decreasing continuously to undetectable frequencies, as might be expected under clonal interference by non-mucoid adaptive genotypes, mucoids fluctuated within detectable frequencies ranges to the end of MyxoEE-6, an outcome suggestive of negatively frequency-dependent selection.

In light of the above results, we designed an experiment to test for such NFD between a mucoid and a non-mucoid clone from population ME4 (cycle 18) resulting from immediate effects of prey-type frequency on prey-type fitness that would be observable over the course of a single cycle of MyxoEE-6 community growth. Such single-cycle NFD could result, for example, if non-mucoids can cheat on mucoids with respect to EPS production [47], not incurring the cost of excess EPS production borne by mucoids but receiving some benefit of such production by neighboring mucoid cells with respect to predation susceptibility. However, this experiment did not detect such NFD between these two focal prey clones, despite confirming a general fitness advantage to the mucoid clone specific to the presence of predation pressure (Figure 2). This outcome might suggest that mucoid vs. non-mucoid fitness is in fact not generally dependent on frequency in a predation-dependent manner, but it is premature to exclude the NFD hypothesis absent further experiments. It remains possible that the observed fitness relationships between the particular ME4 clones chosen for this experiment are not representative of those between most mucoid and non-mucoid genotypes and/or that details of the experimental conditions under which the competition experiment was performed (e.g., starting predator density) prevent the manifestation of NFD that did actually occur during MyxoEE-6. Alternatively, the productivity cost of mucoidy demonstrated here (Figures 4 and 5c) suggested to us the possibility of a distinct form of NFD that, if it occurs, would only play out over multiple growth cycles.

We speculate that the productivity cost and predation-resistance of mucoidy have the potential to collectively mediate NFD of mucoid fitness by reducing predator numbers when mucoids are sufficiently frequent in one growth cycle, which may in turn impact mucoid vs. non-mucoid fitness in the subsequent growth cycle. Under MyxoEE-6 conditions, *E. coli* grows much faster than *M. xanthus* and *M. xanthus* is completely dependent on *E. coli* for growth. For these reasons, we hypothesize that *E. coli* populations generally reached carrying capacity within each growth cycle before *M. xanthus* predator populations increased sufficiently to impose strong predation pressure that would selectively favor mucoids. Due to the productivity cost of mucoidy, non-mucoids may have generally outcompeted mucoids upon completion of prey growth earlier in each cycle, while mucoids may have often outcompeted non-mucoids later in each cycle, after predator density had increased sufficiently to impose strong predation pressure and thus favor mucoidy. In this scenario, sufficiently high late-cycle predation pressure promotes the net increase of mucoids over the entire cycle, despite a disadvantage of mucoids upon completion of prey growth before predators increase to large population sizes. However, mucoid frequency in one MyxoEE-6 cycle might impact the degree of late-cycle predation pressure in the next cycle if predators exhibit lower productivity on mucoid cells than non-mucoids.

This latter scenario would require first that predators exhibit lower productivity on mucoids than on non-mucoids, a plausible scenario given that mucoid prey have lower productivity and are less susceptible to predatory death. It would also require that lower predator productivity caused by relatively high mucoid frequency in one cycle decreases total predation pressure over the next cycle sufficiently to give non-mucoids a net whole-cycle advantage over mucoids in that subsequent cycle. This hypothetical scenario involves a negative feedback loop analogous to such loops proposed or demonstrated to maintain diversity in other predator-prey systems [28]. A test of this hypothesis with respect to MyxoEE-6 would require additional experiments to examine whether these requirements are often met across a range of mucoid vs non-mucoid genotype combinations.

Mucoid phenotypes in *E. coli* are associated with increased production of exopolysaccharides, especially colanic acid (CA-EPS) [48,49] and have been linked with mutations in *rcs*, *yjb* and *yrf* genes [40,48]. Given that all sequenced mucoid *E. coli* clones carried mutations in the *yjbEFGH* operon, while none of the non-mucoid clones had such a mutation, it appears that mucoidy in MyxoEE-6 was achieved predominantly by mutation of this operon, which is involved in the synthesis of an uncharacterised exopolysaccharide [39]. Interestingly, deletion of this operon in a Δ*rpoS* background has been shown to result in CA-EPS synthesis and mucoidy [48]. Apart from the mutations in *yjbEFGH*, the mucoid strain from lineage ME4 had an additional mutation upstream of *rcsA*, which is a known positive regulator for colanic acid synthesis [40,50]. Thus, mutations in *yjbEFGH* and *rcsA* are likely to mediate the apparent tradeoff between resistance to *M. xanthus* predation and productivity, although additional genetic experiments would be required to demonstrate this directly. Collectively, mucoidy seems to have been achieved largely through mutating the same operon, but in different genes and sites. These distinct mutations may have themselves generated the phenotypic variation among mucoid genotypes documented here, or they may have interacted with other mutations to generate such diversity.

Natural microbial communities contain myriad species of prey and predators co-existing in complex food chains and webs [51–53]. The shifting balance of selection imposed by predation vs resource utilisation is likely to contribute to this diversity [28,41,43]. Our results with simple synthetic bacterial communities initially derived from just one prey genotype and one predator genotype suggest that both sympatric and allopatric prey diversity that is evolutionarily induced by bacterial predators can be long-lived. This suggests more strongly than previous studies that predators of microbes and myxobacteria in particular promote greater intra-specific diversity across many prey types over long evolutionary time scales than would co-exist in the absence of predation. Future experiments with more complex communities are required to test this prediction.

At the inter-specific level, it is now well understood that variable levels and forms of intraspecific diversity impact community evolution (e.g., [54–56]). Thus, myxobacteria are likely to shape microbial community composition and structure both directly through differential predation of distinct species [21] and indirectly by shaping long-term patterns of intraspecific diversity that in turn influence inter-specific fitness relationships.

Supplementary Materials: The following are available online at https://www.mdpi.com/article/10.3390/microorganisms9102079/s1, Figure S1: Mucoid-frequency estimates for all evolved MyxoEE-6 *E. coli* populations after cycles 18 and 25.

Author Contributions: R.R.N. and G.J.V. designed the study and analysed the data. R.R.N. conducted the experiments. R.R.N. wrote the initial draft. Review and editing by G.J.V. and R.R.N. All authors have read and agreed to the published version of the manuscript.

Funding: This research received no external funding.

Institutional Review Board Statement: Not applicable.

Informed Consent Statement: Not applicable.

Data Availability Statement: Raw data presented in figures available at Figshare (doi:10.6084/m9.figshare.16726465).

Acknowledgments: We thank Dominique Schneider, Alex Hall and Sébastien Wielgoss for helpful discussions and inputs, Marco La Fortezza for help with microscopy and Lisa Freund for experimental assistance.

Conflicts of Interest: The authors declare no conflict of interest.

References

1. Estes, J.A.; Crooks, K.; Holt, R.D. Predators, Ecological Role of. In *Encyclopedia of Biodiversity*, 2nd ed.; Academic Press: Waltham, MA, USA, 2013; pp. 229–249. [CrossRef]
2. Viola, D.V.; Mordecai, E.A.; Jaramillo, A.G.; Sistla, S.A.; Albertson, L.K.; Gosnell, J.S.; Cardinale, B.J.; Levine, J.M. Competition-defense tradeoffs and the maintenance of plant diversity. *Proc. Natl. Acad. Sci. USA* **2010**, *107*, 17217–17222. [CrossRef]
3. Krediet, C.J.; Donahue, M.J. Growth-mortality trade-offs along a depth gradient in *Cancer borealis*. *J. Exp. Mar. Biol. Ecol.* **2009**, *373*, 133–139. [CrossRef]
4. Cadotte, M.W.; Fortner, A.M.; Fukami, T. The effects of resource enrichment, dispersal, and predation on local and metacommunity structure. *Oecologia* **2006**, *149*, 150–157. [CrossRef]
5. Sherr, E.B.; Sherr, B.F. Significance of predation by protists in aquatic microbial food webs. *Antonie Van Leeuwenhoek* **2002**, *81*, 293–308. [CrossRef]
6. Rodríguez-Zaragoza, S. Ecology of Free-Living Amoebae. *Crit. Rev. Microbiol.* **1994**, *20*, 225–241. [CrossRef] [PubMed]
7. Shtonda, B.B.; Avery, L. Dietary choice behavior in *Caenorhabditis elegans*. *J. Exp. Biol.* **2006**, *209*, 89–102. [CrossRef] [PubMed]
8. Pérez, J.; Moraleda-Muñoz, A.; Marcos-Torres, F.J.; Muñoz-Dorado, J. Bacterial predation: 75 years and counting! *Environ. Microbiol.* **2016**, *18*, 766–779. [CrossRef]
9. Williams, H.N.; Lymperopoulou, D.S.; Athar, R.; Chauhan, A.; Dickerson, T.L.; Chen, H.; Laws, E.; Berhane, T.K.; Flowers, A.R.; Bradley, N.; et al. *Halobacteriovorax*, an underestimated predator on bacteria: Potential impact relative to viruses on bacterial mortality. *ISME J.* **2016**, *10*, 491–499. [CrossRef]
10. Corno, G.; Jurgens, K. Direct and Indirect Effects of Protist Predation on Population Size Structure of a Bacterial Strain with High Phenotypic Plasticity. *Appl. Environ. Microbiol.* **2006**, *72*, 78–86. [CrossRef]
11. Queck, S.Y.; Weitere, M.; Moreno, A.M.; Rice, S.A.; Kjelleberg, S. The role of quorum sensing mediated developmental traits in the resistance of *Serratia marcescens* biofilms against protozoan grazing. *Environ. Microbiol.* **2006**, *8*, 1017–1025. [CrossRef]
12. Müller, S.; Strack, S.N.; Hoefler, B.C.; Straight, P.D.; Kearns, D.B.; Kirby, J.R. Bacillaene and sporulation protect *Bacillus subtilis* from predation by *Myxococcus xanthus*. *Appl. Environ. Microbiol.* **2014**, *80*, 5603–5610. [CrossRef]
13. Mazzola, M.; de Bruijn, I.; Cohen, M.F.; Raaijmakers, J.M. Protozoan-Induced Regulation of Cyclic Lipopeptide Biosynthesis Is an Effective Predation Defense Mechanism for *Pseudomonas fluorescens*. *Appl. Environ. Microbiol.* **2009**, *75*, 6804–6811. [CrossRef] [PubMed]
14. Jousset, A. Ecological and evolutive implications of bacterial defences against predators. *Environ. Microbiol.* **2011**, *14*, 1830–1843. [CrossRef] [PubMed]
15. Scanlan, P.D.; Buckling, A. Co-evolution with lytic phage selects for the mucoid phenotype of *Pseudomonas fluorescens* SBW25. *ISME J.* **2012**, *6*, 1148–1158. [CrossRef]
16. Koskella, B. Phage-Mediated Selection on Microbiota of a Long-Lived Host. *Curr. Biol.* **2013**, *23*, 1256–1260. [CrossRef] [PubMed]
17. Wielgoss, S.; Bergmiller, T.; Bischofberger, A.M.; Hall, A.R. Adaptation to Parasites and Costs of Parasite Resistance in Mutator and Nonmutator Bacteria. *Mol. Biol. Evol.* **2016**, *33*, 770–782. [CrossRef]
18. Muñoz-Dorado, J.; Marcos-Torres, F.J.; García-Bravo, E.; Moraleda-Muñoz, A.; Pérez, J. Myxobacteria: Moving, Killing, Feeding, and Surviving Together. *Front. Microbiol.* **2016**, *7*, 781. [CrossRef]
19. Berleman, J.E.; Zemla, M.; Remis, J.P.; Liu, H.; Davis, A.E.; Worth, A.N.; West, Z.; Zhang, A.; Park, H.; Bosneaga, E.; et al. Exopolysaccharide microchannels direct bacterial motility and organize multicellular behavior. *ISME J.* **2016**, *10*, 2620–2632. [CrossRef]
20. Thiery, S.; Kaimer, C. The Predation Strategy of *Myxococcus xanthus*. *Front. Microbiol.* **2020**, *11*, 2. [CrossRef]
21. Morgan, A.D.; MacLean, R.C.; Hillesland, K.L.; Velicer, G.J. Comparative analysis of *Myxococcus* predation on soil bacteria. *Appl. Environ. Microbiol.* **2010**, *76*, 6920–6927. [CrossRef]

22. Rendueles, O.; Velicer, G.J. Hidden paths to endless forms most wonderful: Complexity of bacterial motility shapes diversification of latent phenotypes. *BMC Evol. Biol.* **2020**, *20*, 145. [CrossRef]
23. Nair, R.R.; Vasse, M.; Wielgoss, S.; Sun, L.; Yu, Y.T.N.; Velicer, G.J. Bacterial predator-prey coevolution accelerates genome evolution and selects on virulence-associated prey defences. *Nat. Commun.* **2019**, *10*, 4301. [CrossRef] [PubMed]
24. Piña, S.E.; Mattingly, S.J. The Role of Fluoroquinolones in the Promotion of Alginate Synthesis and Antibiotic Resistance in *Pseudomonas aeruginosa*. *Curr. Microbiol.* **1997**, *35*, 103–108. [CrossRef] [PubMed]
25. Miskinyte, M.; Sousa, A.; Ramiro, R.S.; de Sousa, J.A.M.; Kotlinowski, J.; Caramalho, I.; Magalhães, S.; Soares, M.P.; Gordo, I. The Genetic Basis of *Escherichia coli* Pathoadaptation to Macrophages. *PLoS Pathog.* **2013**, *9*, e1003802. [CrossRef] [PubMed]
26. Landau, E.; Shapira, R. Effects of Subinhibitory Concentrations of Menthol on Adaptation, Morphological, and Gene Expression Changes in Enterohemorrhagic *Escherichia coli*. *Appl. Environ. Microbiol.* **2012**, *78*, 5361–5367. [CrossRef]
27. Paine, R.T. The Pisaster-Tegula Interaction: Prey Patches, Predator Food Preference, and Intertidal Community Structure. *Ecology* **1969**, *50*, 950–961. [CrossRef]
28. Soto-Ortiz, L. The Regulation of Ecological Communities Through Feedback Loops: A Review. *Res. Zool.* **2015**, *5*, 1–15. [CrossRef]
29. Menge, B.A.; Iles, A.C.; Freidenburg, T.L. Keystone Species. In *Encyclopedia of Biodiversity*, 2nd ed.; Academic Press: Waltham, MA, USA, 2013; pp. 442–457. [CrossRef]
30. Bond, A.B.; Kamil, A.C. Apostatic selection by blue jays produces balanced polymorphism in virtual prey. *Nature* **1998**, *395*, 594–596. [CrossRef]
31. Good, B.H.; McDonald, M.J.; Barrick, J.E.; Lenski, R.E.; Desai, M.M. The dynamics of molecular evolution over 60,000 generations. *Nature* **2017**, *551*, 45–50. [CrossRef]
32. Lang, G.I.; Rice, D.P.; Hickman, M.J.; Sodergren, E.; Weinstock, G.M.; Botstein, D.; Desai, M.M. Pervasive genetic hitchhiking and clonal interference in forty evolving yeast populations. *Nature* **2013**, *500*, 571–574. [CrossRef] [PubMed]
33. Maddamsetti, R.; Lenski, R.E.; Barrick, J.E. Adaptation, clonal interference, and frequency-dependent interactions in a long-term evolution experiment with *Escherichia coli*. *Genetics* **2015**, *200*, 619–631. [CrossRef]
34. Velicer, G.J. Pleiotropic effects of adaptation to a single carbon source for growth on alternative substrates. *Appl. Environ. Microbiol.* **1999**, *65*, 264–269. [CrossRef]
35. R Core Team. *R: A Language and Environment for Statistical Computing*; R Foundation for Statistical Computing: Vienna, Austria, 2021.
36. Wickham, H. *ggplot2: Elegant Graphics for Data Analysis*; Springer: New York, NY, USA. 2016.
37. Velicer, G.J.; Kroos, L.; Lenski, R.E. Developmental cheating in the social bacterium *Myxococcus xanthus*. *Nature* **2000**, *404*, 598–601. [CrossRef] [PubMed]
38. Fiegna, F.; Velicer, G.J. Competitive fates of bacterial social parasites: Persistence and self-induced extinction of *Myxococcus xanthus* cheaters. *Proc. Biol. Sci. R. Soc.* **2003**, *270*, 1527–1534. [CrossRef] [PubMed]
39. Ferrières, L.; Aslam, S.N.; Cooper, R.M.; Clarke, D.J. The yjbEFGH locus in *Escherichia coli* K-12 is an operon encoding proteins involved in exopolysaccharide production. *Microbiology* **2007**, *153*, 1070–1080. [CrossRef] [PubMed]
40. Majdalani, N.; Gottesman, S. The Rcs phosphorelay: A Complex Signal Transduction System. *Annu. Rev. Microbiol.* **2005**, *59*, 379–405. [CrossRef] [PubMed]
41. Corno, G.; Villiger, J.; Pernthaler, J. Coaggregation in a microbial predator-prey system affects competition and trophic transfer efficiency. *Ecology* **2013**, *94*, 870–881. [CrossRef]
42. Menge, B.A.; Foley, M.M.; Robart, M.J.; Richmond, E.; Noble, M.; Chan, F. Keystone predation: Trait-based or driven by extrinsic processes? Assessment using a comparative-experimental approach. *Ecol. Monogr.* **2021**, *91*, e01436. [CrossRef]
43. Vance, R.R. Predation and Resource Partitioning in One Predator—Two Prey Model Communities. *Am. Nat.* **1978**, *112*, 797–813. [CrossRef]
44. Estévez, D.; Galindo, J.; Rolán-Alvarez, E. Negative frequency-dependent selection maintains shell banding polymorphisms in two marine snails (*Littorina fabalis* and *Littorina saxatilis*). *Ecol. Evol.* **2021**, *11*, 6381–6390. [CrossRef]
45. Weir, J.C. Search Image-Driven Apostatic Selection and the Evolution of Phenotypic Polymorphism. *eLS* **2021**, *2*, 1–10. [CrossRef]
46. Barrett, R.D.H.; Laurent, S.; Mallarino, R.; Pfeifer, S.P.; Xu, C.C.Y.; Foll, M.; Wakamatsu, K.; Duke-Cohan, J.S.; Jensen, J.D.; Hoekstra, H.E. Linking a mutation to survival in wild mice. *Science* **2019**, *363*, 499–504. [CrossRef] [PubMed]
47. Dragoš, A.; Lakshmanan, N.; Martin, M.; Horváth, B.; Maróti, G.; García, C.F.; Lieleg, O.; Kovács, T. Evolution of exploitative interactions during diversification in *Bacillus subtilis* biofilms. *FEMS Microbiol. Ecol.* **2018**, *94*, fix155. [CrossRef] [PubMed]
48. Ionescu, M.; Belkin, S. Overproduction of Exopolysaccharides by an *Escherichia coli* K-12 rpoS Mutant in Response to Osmotic Stress. *Appl. Environ. Microbiol.* **2009**, *75*, 483–492. [CrossRef] [PubMed]
49. Ionescu, M.; Belkin, S. Simple quantification of bacterial envelope-associated extracellular materials. *J. Microbiol. Methods* **2009**, *78*, 302–306. [CrossRef] [PubMed]
50. Navasa, N.; Rodríguez-Aparicio, L.; Ferrero, M.; Monteagudo-Mera, A.; Martínez-Blanco, H. Polysialic and colanic acids metabolism in *Escherichia coli* K92 is regulated by RcsA and RcsB. *Biosci. Rep.* **2013**, *33*, 405–415. [CrossRef] [PubMed]
51. Lueders, T.; Kindler, R.; Miltner, A.; Friedrich, M.W.; Kaestner, M. Identification of bacterial micropredators distinctively active in a soil microbial food web. *Appl. Environ. Microbiol.* **2006**, *72*, 5342–5348. [CrossRef]
52. Johnke, J.; Cohen, Y.; de Leeuw, M.; Kushmaro, A.; Jurkevitch, E.; Chatzinotas, A. Multiple micro-predators controlling bacterial communities in the environment. *Curr. Opin. Biotechnol.* **2014**, *27*, 185–190. [CrossRef]

53. Petters, S.; Groß, V.; Söllinger, A.; Pichler, M.; Reinhard, A.; Bengtsson, M.M.; Urich, T. The soil microbial food web revisited: Predatory myxobacteria as keystone taxa? *ISME J.* **2021**, *15*, 2665–2675. [CrossRef]
54. Bolnick, D.I.; Amarasekare, P.; Araújo, M.S.; Bürger, R.; Levine, J.M.; Novak, M.; Rudolf, V.H.; Schreiber, S.J.; Urban, M.C.; Vasseur, D.A. Why intraspecific trait variation matters in community ecology. *Trends Ecol. Evol.* **2011**, *26*, 183–192. [CrossRef]
55. Padfield, D.; Vujakovic, A.; Paterson, S.; Griffiths, R.; Buckling, A.; Hesse, E. Evolution of diversity explains the impact of pre-adaptation of a focal species on the structure of a natural microbial community. *ISME J.* **2020**, *14*, 2877–2889. [CrossRef] [PubMed]
56. Violle, C.; Enquist, B.J.; McGill, B.J.; Jiang, L.; Albert, C.H.; Hulshof, C.; Jung, V.; Messier, J. The return of the variance: Intraspecific variability in community ecology. *Trends Ecol. Evol.* **2012**, *27*, 244–252. [CrossRef] [PubMed]

microorganisms

MDPI

Article

Behavioral Interactions between Bacterivorous Nematodes and Predatory Bacteria in a Synthetic Community

Nicola Mayrhofer [1], Gregory J. Velicer [1], Kaitlin A. Schaal [1,*,†] and Marie Vasse [1,2,*,†]

[1] Institute of Integrative Biology, ETH Zürich, Universitätstrasse 16, 8092 Zürich, Switzerland; nicola.mayrhofer@usys.ethz.ch (N.M.); gregory.velicer@env.ethz.ch (G.J.V.)
[2] MIVEGEC (UMR 5290 CNRS, IRD, UM), CNRS, 34394 Montpellier, France
* Correspondence: kaitlin.schaal@env.ethz.ch (K.A.S.); contact@marievasse.eu (M.V.)
† Shared last authorship and these authors contributed equally to this work.

Abstract: Theory and empirical studies in metazoans predict that apex predators should shape the behavior and ecology of mesopredators and prey at lower trophic levels. Despite the ecological importance of microbial communities, few studies of predatory microbes examine such behavioral res-ponses and the multiplicity of trophic interactions. Here, we sought to assemble a three-level microbial food chain and to test for behavioral interactions between the predatory nematode *Caenorhabditis elegans* and the predatory social bacterium *Myxococcus xanthus* when cultured together with two basal prey bacteria that both predators can eat—*Escherichia coli* and *Flavobacterium johnsoniae*. We found that >90% of *C. elegans* worms failed to interact with *M. xanthus* even when it was the only potential prey species available, whereas most worms were attracted to pure patches of *E. coli* and *F. johnsoniae*. In addition, *M. xanthus* altered nematode predatory behavior on basal prey, repelling *C. elegans* from two-species patches that would be attractive without *M. xanthus*, an effect similar to that of *C. elegans* pathogens. The nematode also influenced the behavior of the bacterial predator: *M. xanthus* increased its predatory swarming rate in response to *C. elegans* in a manner dependent both on basal-prey identity and on worm density. Our results suggest that *M. xanthus* is an unattractive prey for some soil nematodes and is actively avoided when other prey are available. Most broadly, we found that nematode and bacterial predators mutually influence one another's predatory behavior, with likely consequences for coevolution within complex microbial food webs.

Keywords: microbial food web; trophic interactions; predator–prey interactions; mesopredator; social bacteria; nematodes; experimental community; behavior

Citation: Mayrhofer, N.; Velicer, G.J.; Schaal, K.A.; Vasse, M. Behavioral Interactions between Bacterivorous Nematodes and Predatory Bacteria in a Synthetic Community. *Microorganisms* **2021**, *9*, 1362. https://doi.org/10.3390/microorganisms9071362

Academic Editor: David Whitworth

Received: 12 March 2021
Accepted: 12 June 2021
Published: 23 June 2021

1. Introduction

Predation is an ancient biological interaction that influences ecosystem resource turnover [1,2] as well as species abundance, diversity, and evolution [3–8]. Predators can be found at all biological scales and include organisms as different as white sharks and microbes. While less familiar, small predators such as protists, nematode worms, and bacteria make fundamental contributions to global biogeochemical cycling [9,10] and are proposed to be key players for both agriculture [11] and human health [12,13]. In addition, predation in microbial communities may have been a driving force in some of the major transitions in evolution, including the origin of the eukaryotic cell [14–17] and the advent of multicellularity [18,19].

The predatory interactions that link members of a community can be represented as food webs, trophic networks that display the flow of energy among community members. Food webs have long been a central concept in ecology and are powerful tools for investigating community structure, the nature and strength of pairwise interactions, and the indirect effects of interactions on various aspects of community ecology. Food web research typically relies on direct observation of organism behavior, but such direct observations are difficult or impossible when studying microbes (especially given that many microbes cannot be cultured under laboratory conditions). To study species' interactions within microbial food webs [20], ecologists rely on computational studies, mathematical modelling, and experiments in simplified microbial systems. Examples include flux-balance analysis [21–23], study of pairwise interactions, and examination of growth and death curves in small communities [24]. These approaches usually adopt a bottom-up strategy, inferring features of the community based on the careful investigation of its components. However, they likely miss out on higher-order interactions and complex behavioral responses. Since predatory interactions may occur in complex networks, involve various partners, and fluctuate over time, these bottom-up approaches might be insufficient to understand the dynamics and broader ecological impacts of microbial predation [25–27].

Despite the ecological importance of microbes [10], microbial predators have only recently received substantial recognition as agents that influence biodiversity by controlling and shaping bacterial communities [9,28–32]. Microbial predators use a wide range of strategies to kill and consume their prey, from the periplasm-invasion strategy of *Bdellovibrio bacteriovorus*, which grows and divides within its prey, to total engulfment by protists and far-range killing by *Streptomyces* species [33]. Cells of *Myxococcus xanthus*, the most studied myxobacteria species, forage in groups, repeatedly reversing direction while attacking and consuming prey [34]. Myxobacteria are predicted to play major roles in shaping the structure and evolution of soil communities due to their ability to eat diverse species as prey [35] and strongly influence prey evolution [36], as well as their abundance in soils [37].

According to current understanding, *M. xanthus* secretes extracellular hydrolytic enzymes that break down prey macromolecules and allow uptake of the released nutrients [38]. Studies of *M. xanthus* predation have examined its molecular mechanisms [34,39,40], the effects of ecological conditions [35,41–43], and (co)evolution with a single prey species [36,44]. Little is understood about how *M. xanthus* may itself be exposed to predation pressure and how it interacts with its own predators. Natural bacterial communities are often grazed by bacteriophagous microfauna such as nematodes and protozoa, which can influence their structure and composition [11,45]. It is very likely that some such bacte-riophagous organisms prey upon *M. xanthus* in natural environments, thus making *M. xanthus* a potential mesopredator, defined as an organism in a given food web that obtains nutrients by killing and consuming other organisms and faces predation risk from larger organisms [46]. In fact, Dahl et al. [47] showed that the predatory nematode *Caenorhabditis elegans* will ingest *M. xanthus* in some contexts. However, whether *C. elegans* achieves net growth from nutrients derived from wild-type *M. xanthus* remains uncertain.

It is not known to what extent the community-ecology effects of microbial predators mirror those of multicellular predators (but see Steffan et al. [48]). In large organisms, intraguild predation (when apex predator and mesopredator also compete for the same basal prey organism) can have direct effects on mesopredator survival and distribution [49]. Ritchie and Johnson [50] reviewed the effects of apex predators on mesopredators and their prey in 94 animal studies and found that, on average, increasing apex predator population size two-fold reduces mesopredator abundance by approximately four-fold. Such effects may be similarly important in communities of microbes. In addition to their direct demographic effects, apex predators can generate substantial behavioral modifications in mesopredators, altering their habitat use and changing their foraging activity, thereby indirectly affecting their survival and growth [50]. It is unclear whether these communities show hierarchical trophic interactions, and, if so, whether bacteriophagous organisms function as apex predators. The study of microbial community dynamics, therefore, requires a better understanding of the direct and indirect interactions between bacterial prey, bacterial predators (potential mesopredators), and bacteriophagous nematodes and protozoa (potential apex predators).

Here, we examined behavioral interactions between two predator species in a synthetic community and investigated how such interactions modulate food web structure. We designed the community to have three trophic levels. Predicted trophic interactions are depicted in Figure 1. We hypothesized *C. elegans* to function as a potential apex predator,

M. xanthus as a potential mesopredator, and *Escherichia coli* and *Flavobacterium johnsoniae* as two basal prey bacteria. We first tested for effects of the bacterial predator on nematode predatory behavior by asking whether (1) *M. xanthus* attracts or repels *C. elegans* in the absence of other prey, (2) bacterial cell death or strain motility alters any effect of *M. xanthus* on *C. elegans*, (3) *M. xanthus* is more or less attractive to *C. elegans* as potential prey than the two basal prey species, and (4) the presence of *M. xanthus* in mixture with one basal prey species in a given prey patch alters its attractiveness to worms. We then asked whether nematodes reciprocally influence *M. xanthus* behavior—specifically, swarming behavior within patches of basal prey—whether due to direct interactions between the predator species or indirectly due to nematode effects on basal prey populations.

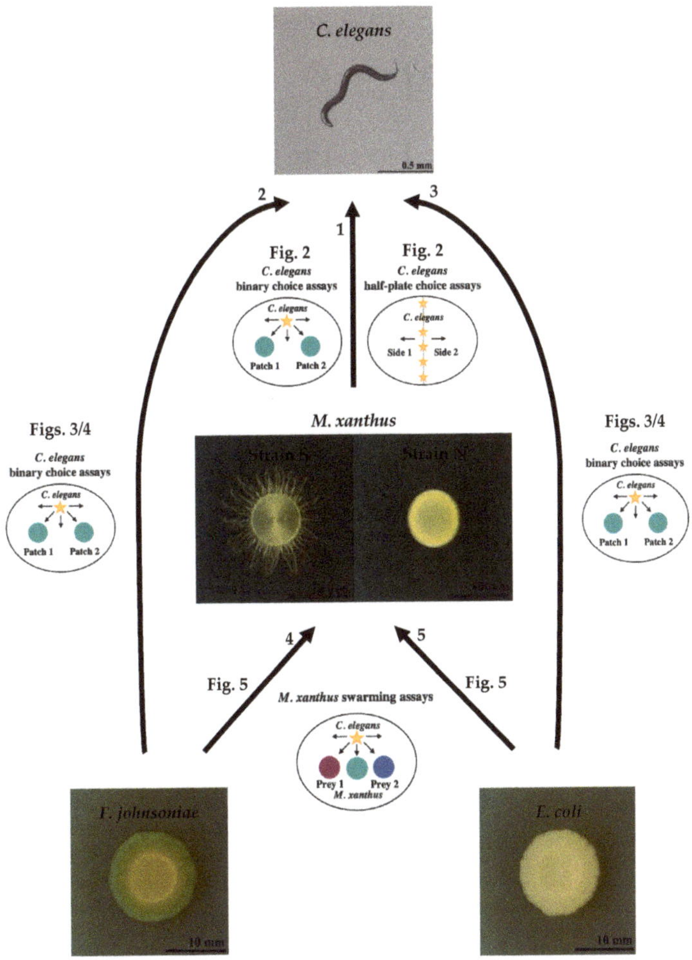

Figure 1. Predicted trophic interactions of the synthetic community and illustrations of experimental designs. We predicted that *C. elegans* might function as a potential apex predator in this food web, preying on all other members of the synthetic community (arrows 1, 2, and 3). We predicted *M. xanthus* to function as a mesopredator that preys upon both basal prey species (arrows 4 and 5) and to experience predation pressure from the nematode apex predator (arrow 1). We show illustrations of the experimental designs that were used to test each depicted interaction and reference the figures that report the relevant results.

2. Materials and Methods

2.1. Bacterial and Nematode Strains

As the hypothesized apex predator, we used *C. elegans* strain N2 (CGC). As hypothesized mesopredators, we used two strains of *Myxococcus xanthus*, GJV1 and GJV71. *M. xanthus* uses two distinct motility systems to drive swarming across solid surfaces, traditionally referred to as the 'A motility system' and the 'S motility system' [51]. Strain GJV1 possesses both systems functionally intact and is a clone of DK1622 [52], which was used in the only prior study reporting interactions between *M. xanthus* and *C. elegans* [47]. For purposes of this paper, we hereafter refer to GJV1 as strain S, for 'swarming'. GJV71 is a nonmotile mutant of GJV1 with major deletions in two genes, one gene essential for A motility (*cglB*) and one gene essential for S motility (*pilA*). GJV71 was referred to as strain 'A1 *cglB*′ in Velicer and Yu [53]. We hereafter refer to GJV71 as strain N, for 'non-swar-ming'. We selected the *Escherichia coli* strain OP50 [54] (CGC, Caenorhabditis Genetic Center) and *Flavobacterium johnsoniae* (ATCC® 17061™) as basal prey bacteria because they represent, respectively, high- and intermediate-quality food sources for *M. xanthus*, promoting *M. xanthus* swarming and growth to different degrees [35,43]. *E. coli* strain OP50 is the standard prey for laboratory populations of *C. elegans* [55]. In preliminary experiments, *F. johnsoniae* sometimes displayed a phenotype with low gliding ability, which inhibited *M. xanthus* predation. In all following experiments, the source of *F. johnsoniae* was a frozen stock originating from a single colony, which we isolated from a normally sprea-ding population.

2.2. Standard Culture Conditions

Unless otherwise indicated, organisms were cultured on 6-cm diameter petri dishes each with 14 mL of 1.5% agar CFcc medium ('clone fruiting' medium [56] supplemented with 1 mM $CaCl_2$ and 0.005 mg/mL cholesterol).

2.3. Culturing C. elegans

We froze *C. elegans* in a 10% DMSO solution and thawed it in minimal salts buffer (M9) with glutamine, according to Pires da Silva et al. [57]. We maintained the worms at room temperature on 1.5% agar nematode growth medium (NGM) dishes seeded with *E. coli* OP50, transferring weekly. We synchronized the life stages of all *C. elegans* populations prior to use in an experiment. Seven days before the start of the experiment, we transferred a small inoculum from a growing population to seeded 1.5% agar high growth medium (HGM) dishes [58]. After 6 days of incubation at room temperature, we washed the agar surface with M9 to collect the worms. We centrifuged them at 173× *g* for 1 min and removed all but 1 mL of supernatant. We then added 3 mL of bleaching solution (6 mL ddH_2O, 6 mL NaOCl (5% Cl), 2 mL 1M NaOH) and waited up to 6 min, vortexing every 2 min. This step dissolved the bodies of the adult worms, releasing the eggs. We then washed the released eggs four times by centrifuging, removing all but 500 µL of supernatant, and adding ddH_2O to 5 mL. After the final wash, we added 3.5 mL of M9 and transferred the egg suspension to a 6-cm petri dish to hatch at room temperature overnight. To prevent contamination, we added 40 µg/mL gentamicin. The next day, we collected the hatched L1 larvae by centrifuging and resuspended them in CFcc liquid. We determined the worm density in the suspension by plating three 1-µL drops on unseeded CFcc agar plates and counting the worms in each drop.

2.4. Culturing M. xanthus

We inoculated *M. xanthus* from freezer stock onto 1.5% agar CTT (10 g/L Casitone, 10 mM Tris pH 8.0, 8 mM $MgSO_4$, 1 mM KPO_4 [59]) dishes and incubated it at 32 °C and 90% relative humidity (rH) for 4–5 days. We then sampled the outer edge of the resulting colony and transferred the inoculum into CTT liquid, shaking at 32 °C and 300 rpm for 1 day until the cultures reached mid-exponential phase, then adjusted them to an absor-bance (OD_{600}) of 5 in CFcc liquid.

2.5. Culturing Prey Bacteria

We streaked *F. johnsoniae* and *E. coli* from freezer stock onto 1.5% agar lysogeny broth (LB, Sigma, St. Louis, MI, USA) dishes and incubated them at 32 °C and 90% rH for 3 days. We transferred single colonies into LB liquid, shaking at 32 °C and 300 rpm for 1 day (or ~10 h in the case of *E. coli*), then adjusted the cultures to an OD_{600} of 5 in CFcc liquid.

2.6. C. elegans' Binary Choice Assays

We inoculated two 15-µL bacteria spots 2 cm apart on CFcc agar and incubated at 25 °C and 50% rH overnight before adding *C. elegans*. We added 20 *C. elegans* L1 larvae suspended in CFcc liquid to the dishes and incubated them at 25 °C and 50% rH. We counted the number of worms in different regions of the petri dish under a dissecting microscope at several time points. For the assay reported in Figure 2B, we observed the worms after 1 h and 18 h, as preliminary experiments indicated that the worm location did not change after 18 h. For the assays reported in Figures 3 and 4, we used the same time points plus an additional one at 25 h to allow the prey and *M. xanthus* to interact and this interaction to potentially affect the worm location. For the assays reported in Figure 2A,B, we prepared both living and dead cultures of *M. xanthus* strains S and N. To kill *M. xanthus*, we resuspended a growing culture to OD 5 in CFcc liquid and incubated at 50 °C for 3 h. We left live cells shaking at 32 °C and 300 rpm during this time, before adjusting the OD. An illustration of the *C. elegans'* binary choice assay design is depicted in Figure 1. Results from these assays are shown in Figures 2A,B, 3 and 4.

2.7. C. elegans Half-Plate Choice Assays

We drew center lines on CFcc plates and prepared both living and dead cultures of *M. xanthus* strains S and N as reported above. We inoculated one half of each petri dish with 100 µL of one of the bacterial cultures or buffer control spread with a 10-µm loop and allowed the inoculum to dry. We bleached *C. elegans* and adjusted the egg suspension to 50 eggs/µL by counting the number of eggs in three 0.5-µL drops. We immediately added 20 µL of egg suspension (approximately 1000 eggs) to each dish along the center line. We incubated the dishes at 25 °C and 50% rH and counted the number of worms on each side of the dish under a dissecting microscope at several time points. We chose the time points of 18 and 42 h to be similar to the time points for the experiment in Figure 2B but with a delay to allow the nematode eggs to hatch on the dish. The plating design of the *C. elegans* half-plate choice assays is illustrated in Figure 1. Results from these assays are shown in Figure 2C,D.

2.8. M. xanthus Swarming Assays

We marked CFcc plates with reference lines and scale bars for image analysis. We inoculated 15 µL each of *F. johnsoniae*, *E. coli*, and *M. xanthus* strain S in a row, with *M. xanthus* in the middle and 1-cm distance between each inoculation spot, and incubated the dishes at 20 °C and 50% rH. We prepared *C. elegans* worms by bleaching them on the same day that we plated the bacteria, and the next day we added the appropriate number of worms to each dish either by manually picking the desired number of individual L1 larvae and adding them directly or by adding 10 µL of worms suspended in CFcc liquid adjusted to the appropriate concentration. We took pictures of the experimental plates every 24 h, and we measured the distance *M. xanthus* swarmed into each prey patch over time with image analysis using Fiji [60]. Figure 1 shows an illustration of the *M. xanthus* swarming assays. Results from these assays are shown in Figure 5.

2.9. Statistical Analysis

We performed all data analysis and statistical testing using R version 3.6.2 and RStudio version 1.2.5033 [61,62]. We tested the effect of *C. elegans* on *M. xanthus* swarming distance on prey using a mixed linear model with prey type (*E. coli* or *F. johnsoniae*) and worm treatment (factor presence/absence in one experiment, continuous variable number of worms in another) as fixed effects. As we measured the swarming distance on the two prey species from the same experimental petri dish, we included the dish identity as a random factor to account for repeated measures. We compared treatment modalities u-sing the Tukey method for multiple comparisons from the emmeans package version 1.4.3 [63]. To evaluate *C. elegans'* choice in the binary choice and half-plate choice assays, we calculated a choice index as in Moore et al. [64]:

$$(\#\text{ worms on side A} - \#\text{ worms on side B})/(\#\text{ worms on side A} + \#\text{ worms on side B}). \quad (1)$$

Null values indicate that the worms did not prefer one side over the other (or that they all left the dish). To compare their attraction to or avoidance of live and dead *M. xanthus* (strain S or N), we used an ANOVA with the options treatment and time as fixed effects. We considered time as a fixed effect because we were interested in whether the differences between the option treatments would change over time. We performed post hoc comparisons with the Tukey method. We further tested whether the worms preferred one option over the others with one-sample *t*-tests against 0 with Bonferroni correction for multiple testing.

3. Results

3.1. Only a Few C. elegans Worms Interact with M. xanthus Regardless of Whether It Is Alive or Dead

To test for interactions between *M. xanthus* and *C. elegans*, we co-cultured worms and bacteria on agar petri dishes in a variety of assays. In all of these assays, the worms had three spatial areas to choose among: (1) outside the assay plate, which could be reached by worms climbing out of the petri dish; (2) agar-surface regions with no bacteria present; and (3) agar-surface areas covered by bacterial cells, with some of these areas being circular patches (binary choice assays) and others covering half of a petri dish (half-plate choice assay).

In our first experiments, only *M. xanthus* was available as potential prey, and we offered both live and dead *M. xanthus* cells to *C. elegans* to test for any effect of cell death on their attractiveness. If *M. xanthus* is an attractive prey item for *C. elegans*, it might attract worms equally whether alive or dead. Alternatively, if *C. elegans* avoids living *M. xanthus* cells because they produce a repellant compound, dead *M. xanthus* might, nonetheless, serve as a palatable food source. Two strains of *M. xanthus* were offered to *C. elegans*, one motile (strain S) and one non-motile (strain N).

In a binary choice assay, we inoculated two circular patches of *M. xanthus* on an agar surface and added L1 larval worms to a bacteria-free region of the plate, equidistant from the two *M. xanthus* patches. In this assay, we counted how many worms left the plate vs. remained on the plate after 1 and 18 h and, of those that remained, how many entered one or the other of the *M. xanthus* patches. In most replicates, regardless of the options provided, large majorities of the worm populations emigrated from the dish (>75% on average), and there was no general difference in the rate of emigration as a function of *M. xanthus* strain identity (ANOVA bacterial identity factor $F_{2,12} = 0.8$, $p = 0.5$, Figure 2A). Those who stayed demonstrated no clear general preference between live vs. dead patches across both strains and both examined time points (1 and 18 h, ANOVA choice:time interaction $F_{2,24} = 5.67$, $p < 0.01$, post hoc Tukey HSD tests $p > 0.2$; Figure 2B and Figure S1A). One exception to the general lack of a strong effect of *M. xanthus* death occurred on dishes containing the motility mutant strain N. On these plates, the worms seemed to initially prefer the live strain N patch after 1 h but then changed their preference to the dead patch by 18 h (post hoc Tukey HSD test $p = 0.01$). Because no similar pattern was seen for strain S, this result suggested that the effect of death on the attractiveness of bacterial cells to nematode predators can vary across conspecific genotypes.

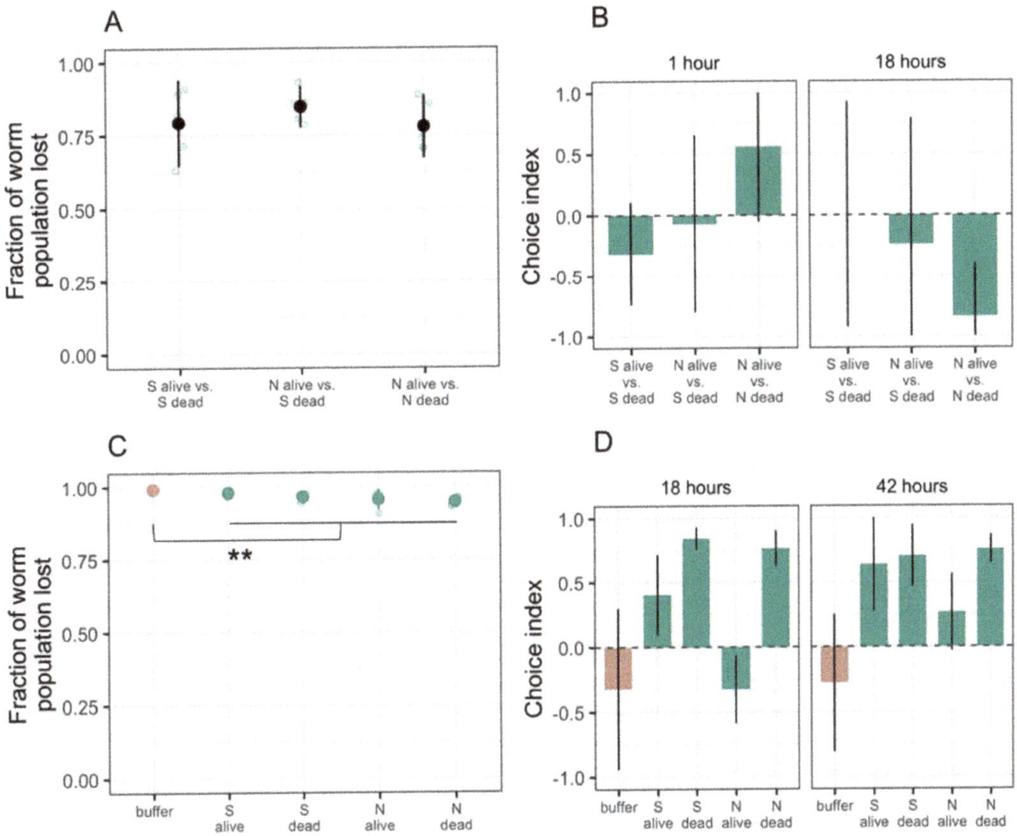

Figure 2. ***C. elegans* prefers *M. xanthus* to buffered agar.** We tested potential effects of living or dead *M. xanthus* on the position of worms presented with choices between either two circular patches of *M. xanthus* (**A**,**B**) or between a half-plate lawn of *M. xanthus* versus a half plate of bacteria-free agar (**C**,**D**). For both experiments, we report the mean fraction of *C. elegans* populations that left the plate (**A**,**C**) and the choices made by the worms that remained on the plate (**B**,**D**). In panel B, positive vs. negative values reflect attraction to the first- vs. second-listed type of bacteria, respectively. In panel D, positive vs. negative values reflect attraction to the inoculated vs. uninoculated half of the plate, respectively. Each large dot is the mean of five biological replicates (shown as transparent blue dots). Error bars represent 95% confidence intervals. In some instances (**B**), 95% confidence intervals extend outside the range of what is biologically possible, [−1,1], so we restricted them to reflect the biological reality. ** $p < 0.01$.

The preference of the worms to leave the experimental petri dishes suggested that *M. xanthus* might repel *C. elegans*. To test this hypothesis, we inoculated *M. xanthus* alone—either strain S or N, alive or dead—onto half of the agar surface of the petri dish, leaving the other half uninoculated. In the previous binary choice assay, the patches of *M. xanthus* were small relative to the agar surface of the experimental petri dishes, reducing the likelihood of finding worms in a patch unless the bacteria actively attracted them. In contrast, in this half-plate choice assay we expected to find 50% of the worms on the plate located in the bacterial lawn, assuming no interactions between the two organisms. We included plates inoculated with sterile resuspension buffer to control for the potential attractive or repulsive effect of the buffer itself. We added *C. elegans* eggs to the midline of the dish and counted how many worms left vs. remained on the plate after hatching and, of those that remained, how many went to each half of the dish.

The overwhelming majority of worms (>90%) preferred to leave the dish both in the absence and presence of *M. xanthus* (Figure 2C). However, a larger proportion of worms remained on the plate when *M. xanthus* was present (~4% on average across all treatments with *M. xanthus*) than on control plates with buffer alone (~1%, $t_{22} = 6.42$, $p < 0.01$). Moreover, the worms that remained were more likely to be located on the inoculated side of the dish on plates with *M. xanthus* versus on plates with only buffer (ANOVA bacterial presence $F_{2,19} = 27.43$, $p < 0.01$, post hoc Tukey HSD test for preference of strain S or strain N over buffer, p-values < 0.01; Figure 2D and Figure S1B). We saw no difference between live and dead (post hoc Tukey HSD test $p = 0.24$) or motile and non-motile *M. xanthus* (post hoc Tukey HSD test $p = 0.44$). Our results revealed intrapopulation heterogeneity in whether worms remain on plates containing only *M. xanthus* and suggested that, among the minority of worms that did remain, *C. elegans* was attracted to *M. xanthus*.

3.2. *C. elegans Prefers Both Basal Prey Species over M. xanthus*

To investigate potential behavioral responses of *C. elegans* to *M. xanthus* relative to other potential prey, we performed additional binary choice assays. We inoculated two patches of bacteria on each dish, allowing the worms to choose between them, remain in the open agar, or emigrate from the dish, and then counted the number of worms in each plate area after 1, 17, and 25 h. In these experiments, each patch contained either one basal prey species alone, one basal prey mixed 1:1 with *M. xanthus* strain S or N, or *M. xanthus* strain S or N alone. We added 20 worms to a bacteria-free region of the plate, allowing them to explore across the agar surface and seek out their preferred prey.

As expected from the previous results (Figure 2A,C), when *M. xanthus* was the only option, most worms had either left the dish entirely or were located in the open agar after 25 h (Figure 3A and Figure S2A). When *C. elegans* could choose between a patch of basal prey mixed with *M. xanthus* and a patch of the same prey without *M. xanthus*, the worms almost invariably chose the latter (ANOVA $F_{13,28} = 10.01$ $p < 0.01$, p-values < 0.05 except for *F. johnsoniae* vs. *F. johnsoniae* + strain S; Figure 3B and Figure S2A). In the absence of *M. xanthus*, *C. elegans* preferred *E. coli* over *F. johnsoniae* (one-sample t-test for choice index < 0 $t_2 = -3.47$ $p = 0.037$; Figures 3C and 4, Figures S2B and S3). However, the presence of *M. xanthus* in *E. coli* patches altered the preference of the worms; *C. elegans* tended to prefer patches of *F. johnsoniae* over mixed patches containing both *E. coli* and *M. xanthus* ($p = 0.096$ and $p = 0.028$ for mixes with strain S and strain N, respectively, 14 two-sided t-tests with Bonferroni–Holm correction; Figures 3C and 4, Figure S2B). In general, the presence of *M. xanthus* drastically reduced patch attractiveness for *C. elegans* at every time point, independent of *M. xanthus* strain identity ($p < 0.1$, t-tests as described above; Figures 3 and 4, Figures S2 and S4).

Depletion of prey patches by *M. xanthus* typically requires many hours or even several days, depending on the prey type [42,43]. In this experiment, many worms localized to either Patch 1 or Patch 2 already after one hour, and by 17 h nearly all did so (among the ones remaining on the dish, Figure S4). We considered it unlikely *M. xanthus* had by that time fully cleared the prey bacteria from the mixed patches. However, to confirm that *C. elegans* did not change its behavior toward the basal prey simply as a consequence of *M. xanthus* having consumed all basal prey within the patch, leaving none to tempt *C. elegans*, we streaked samples of each mixed patch on LB petri dishes at the end of the experiment to check for the presence of *E. coli* or *F. johnsoniae*. In seven out of 24 cases, we could verify that the prey bacterium was still present in the mixed patch after 25 h (Table S1). Even in these cases, the worms tended not to go into the mixed patches, suggesting that the presence of *M. xanthus* in mixed bacterial patches repelled *C. elegans* toward pure patches of *E. coli* or *F. johnsoniae*.

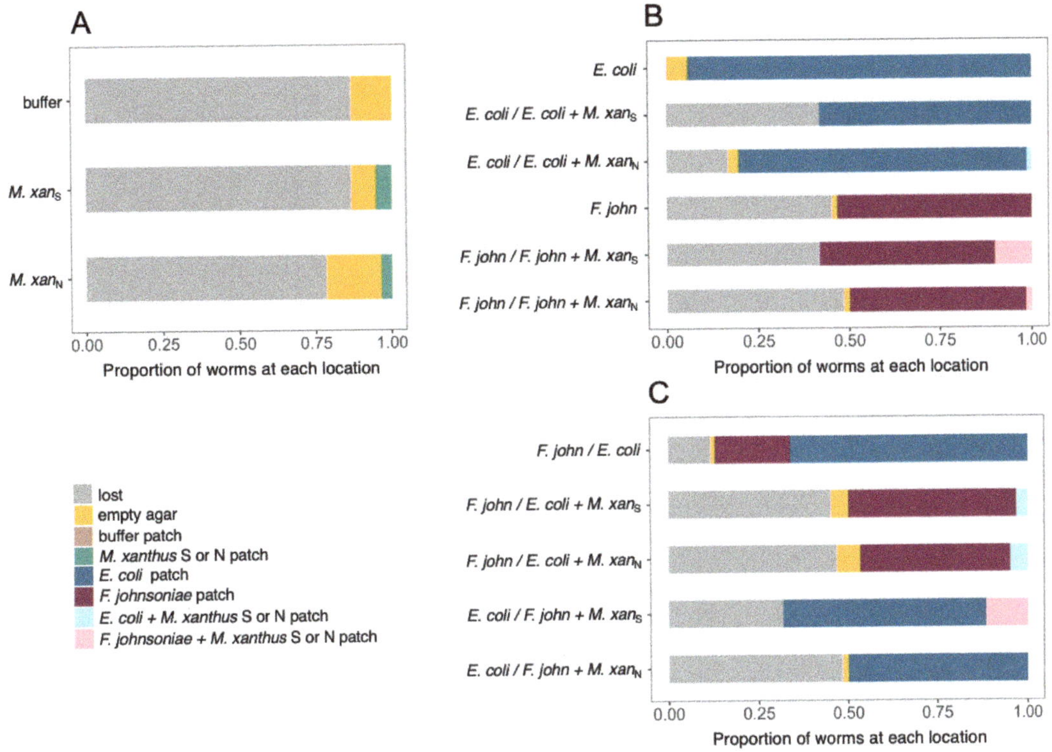

Figure 3. ***C. elegans*** **prefers prey patches not containing** ***M. xanthus*****.** Here we show worm locations relative to two circular prey patches after 25 h on plates where either (**A**) the patches contained either buffer or *M. xanthus*, (**B**) one patch contained *E. coli* or *F. johnsoniae* and the other patch contained the same with or without pre-mixed *M. xanthus*, or (**C**) one patch contained *E. coli* or *F. johnsoniae* and the other patch contained the other prey species with or without pre-mixed *M. xanthus*. Worms found on the plate but not in a patch are indicated in yellow, and worms that had left the plate by 25 h are indicated in grey. Bars are mean worm counts from three biological replicates. Raw data are shown in Figure S2. '*F. john*' = *F. johnsoniae*, '$M.xan_N$', '$M.xan_S$' = *M. xanthus* strains N and S, respectively.

3.3. *M. xanthus* Responds Behaviorally to *C. elegans* in a Prey-Dependent Manner

To characterize potential behavioral responses of *M. xanthus* to the presence of *C. elegans*, we performed a binary choice assay similar to those above. In this assay, we added *C. elegans* to an agar petri dish inoculated with two patches of basal prey, one of *E. coli* (e.g., Figure 5A) and one of *F. johnsoniae*, and one patch of *M. xanthus* at the midpoint between the prey such that it would encounter them upon swarming outward. Nematodes might alter *M. xanthus* swarming behavior due to direct interactions with *M. xanthus* or due to indirect effects of resource competition for basal prey, which, in turn, might be affected by the type of basal prey environment.

We measured *M. xanthus* swarming rate in each prey-patch type on dishes with (Figure 5A) and without worms. This swarming-rate measure encompasses both the ability to penetrate the prey patch and predatory performance inside the patch [43]. For the treatment with *C. elegans*, we added 10 worms to a bacteria-free region of the plate. The nematodes had no effect on *M. xanthus* swarming rate in the *F. johnsoniae* patches (ANOVA basal prey identity:worm presence interaction $F_{1,36} = 8.82$ $p < 0.01$, post hoc Tukey HSD test $p = 0.99$; Figure 5B,C) but significantly increased *M. xanthus* swarming in the *E. coli* patches (post hoc Tukey HSD test $p = 0.0004$; Figure 5B,C).

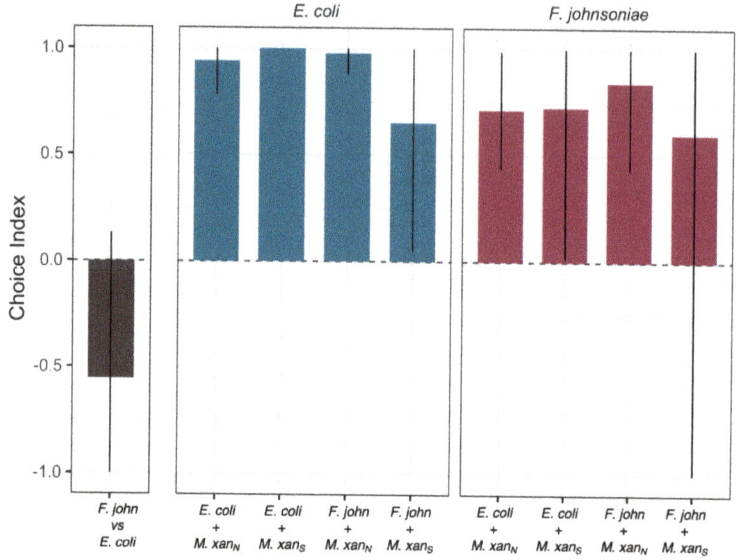

Figure 4. Choice indices for circular patches of basal prey. The leftmost section shows the preference of *C. elegans* for *E. coli* over *F. johnsoniae*. The next two sections show the choice of the worms between a mono-species patch of basal prey (indicated above the panel, *E. coli* = blue or *F. johnsoniae* = red) and a second patch (indicated on the *x*-axis) consisting of basal prey mixed with *M. xanthus* either strain S or strain N. Bars are means from three biological replicates and error bars are 95% confidence intervals. For the choice *E. coli/E. coli* + *M. xan*$_S$ the error bar is zero. In some instances, 95% confidence intervals extend outside the range of what is biologically possible, [−1,1], so we restricted them to reflect the biological reality. '*F. john*' = *F. johnsoniae*, '*M.xan*$_N$', '*M.xan*$_S$' = *M. xanthus* strains N and S, respectively.

The effect of worms in the *E. coli* patches only became visible after day 5 (Figure 5C), and we hypothesized that this could be explained by the onset of worm reproduction, as the worms reached maturity, and subsequent increase in the worm population size. To evaluate whether larger populations of *C. elegans* lead to an increase in *M. xanthus* swarming rate, we repeated the experiment using different numbers of worms. For this experiment, we show the swarming distances between days 3 and 5 (rather than days 5–8 as in the previous assay) to capture the response of *M. xanthus* to the inoculated number of worms rather than to a growing population, as within this time frame the worms had not yet completed their development. We found that *C. elegans* increased *M. xanthus* swarming on *E. coli* to a similar small degree regardless of worm population size (linear model $F_{1,10} = 0.82$, $p = 0.39$, adjusted $R^2 = -0.01$; Figure 5D). It, therefore, remains unclear whether the time delay is simply a delay in *M. xanthus* response or whether it has to do with the developmental progress of the worms. However, in contrast to our first experiment, *C. elegans* clearly increased *M. xanthus* swarming rate on *F. johnsoniae,* but did so only as a function of worm population size (linear $F_{1,10} = 32.01$, $p < 0.001$, adjusted $R^2 = 0.74$; Figure 5D), such that an effect of *C. elegans* was only evident when hundreds of worms were added.

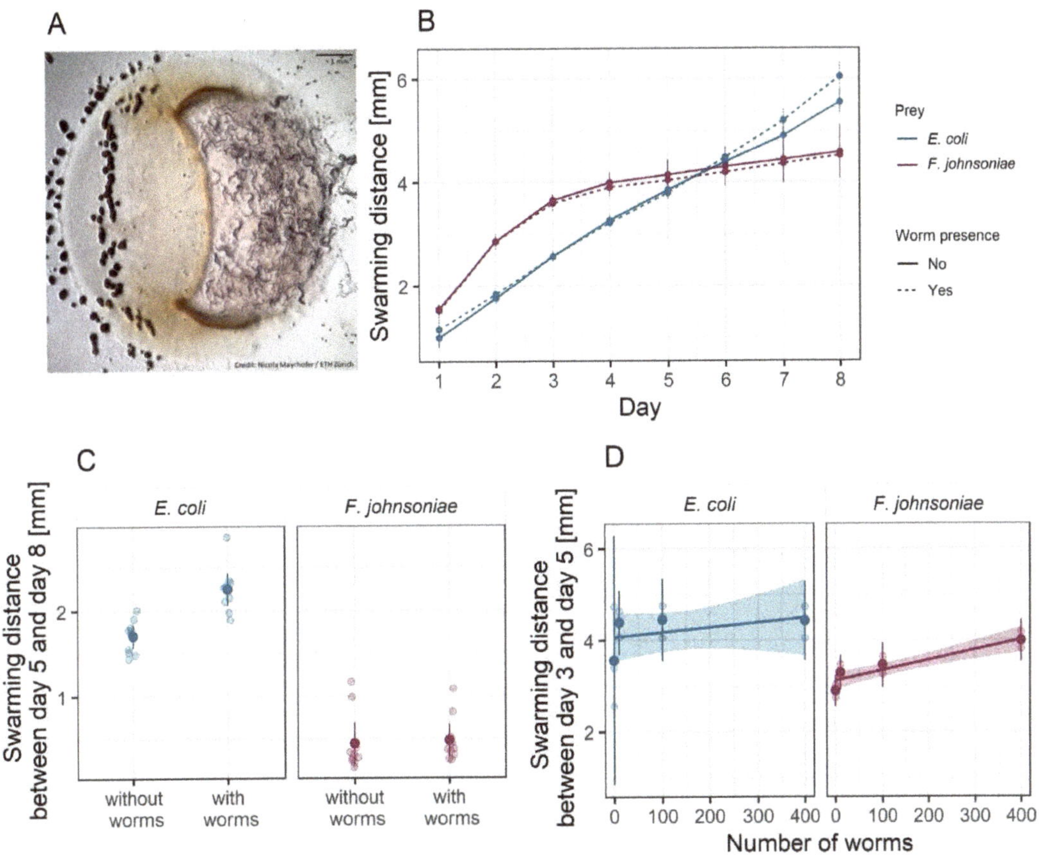

Figure 5. ***C. elegans*** **presence increases** ***M. xanthus*** **predatory swarming rate.** (**A**) Picture of *C. elegans* (right) and *M. xanthus* strain S (orange lawn and black fruiting bodies on the left) preying upon an *E. coli* patch (raised circle). We estimated the predatory performance of *M. xanthus* as the swarming distance along the horizontal midline of the prey patches of *E. coli* (blue) and *F. johnsoniae* (red). We show the swarming distance over time (**B**) in the presence (dotted lines) and absence (solid lines) of *C. elegans'* populations initiated with 10 worms (which started reproducing at day 5), and the associated total swarming distances between days 5 and 8 (**C**) from the same experiment. Panel (**D**) depicts the total swarming distances between days 3 and 5 in the presence of different numbers of *C. elegans* (the worms did not reach maturity until day 5 and so did not reproduce during this experiment). Each large dot is the mean of ten (**B**,**C**) or three (**D**) biological replicates (shown as transparent dots). Error bars and shaded areas represent 95% confidence intervals of the means and the regression lines, respectively.

4. Discussion

Despite their importance in microbial population turnover and community dynamics, there has been little research on microbial trophic chains. Here we investigated interactions between two bacterivorous predators, *M. xanthus* and *C. elegans*, in the context of a synthetic community that included two species of basal prey (*E. coli* and *F. johnsoniae*). We found that *M. xanthus* generally repels *C. elegans* relative to the effects of the two basal prey. When *M. xanthus* was the only prey option available, most worms departed our experimental predation dishes (Figure 2A,C and Figure 3A), whereas when either or both of the basal prey were offered most of the worms remained on the plates (Figure 3B,C). *M. xanthus* was not entirely repulsive to worms though. When only *M. xanthus* was offered, the few worms that remained on the plate localized more frequently within areas with

M. xanthus than on open agar (Figure 2D). However, the presence of *M. xanthus* in a prey patch mixed with a basal prey species repelled *C. elegans* when a separate monoculture patch of either of the basal prey species was available (Figure 3B,C and Figure 4). We further showed that *C. elegans* can alter *M. xanthus* behavior by increasing the bacterial predator's swarming rate across patches of basal prey, and that such behavior alteration depends on basal-prey identity. Our results highlight the importance of predator–predator interactions in microbial communities and the idea that other community members may often considerably modify pairwise behavioral interactions between organisms in such communities.

Theory and previous experiments in metazoans predict that the behavioral interactions between apex predators and mesopredators are driven mainly by the former [65–67]. In our microbial system, however, when a pure patch of basal prey was available, *C. elegans* avoided foraging areas occupied by *M. xanthus* even when they also contained the worms' preferred prey (Figures 3 and 4). One limitation of our study is that we assessed the behavior of the worms based on discrete time points instead of continuous observation. However, the worms' choice of basal prey patches did not change between three different time points, suggesting that most worms remained in the prey patch they had initially entered (Figure S4). Although in this study we did not formally test whether *C. elegans* can use *M. xanthus* as a food source to fuel worm population growth, our results suggest that the behavioral interactions between the two may prevent *C. elegans* from ac-ting as an apex predator in this system. Despite Dahl and colleagues' conclusion that *C. elegans* can be a predator of *M. xanthus* [47], we found that *C. elegans* worms seem to consider *M. xanthus* to be an unpalatable food source. They avoid it whenever more palatable prey is available (Figure 3B,C and Figure 4), and a majority of individuals avoid it even when there is no other prey available (Figure 2A,C and Figure 3A); *C. elegans* does not similarly avoid the basal prey (Figure 3B,C). Still, minorities of worms did interact with *M. xanthus* (Figure 2A,C and Figure 3A) and preferred it over sterile buffer (Figure 2C,D). This could reflect a level of behavioral heterogeneity in the worm population, potentially due to different feeding preferences, predatory behaviors, or sensitivity to repellant compounds produced by *M. xanthus* across individual worms. Such behavior by some worms indicates that there may be some conditions under which *C. elegans* can be attracted to *M. xanthus*, for example, when no other prey source is available.

C. elegans uses its nervous system to recognize different bacteria in its environment [68] and to modify its locomotive behavior in response to prey quality [69]. It can learn to recognize and approach high-quality prey [69] and to avoid pathogens [70]. There is evidence that learned pathogen avoidance is modulated by changes in gene expression that are heritable through four generations [64]. While the mechanistic reasons for the avoidance behaviors we observed here remain to be investigated, these behaviors are consistent with the hypotheses that (i) *M. xanthus*' large secondary metabolome [39,71–73] contains some compounds with a primary or secondary defensive function against predators [74], and (ii) the worms can either sense them at a distance or learn to avoid them after the first encounter [70]. For example, *C. elegans* is known to avoid some *Serratia marcescens* strains after coming into contact with the serrawettin surfactants that the bacteria use for swarming motility [75]. As *M. xanthus* A-motility in particular involves the secretion of a polysaccharide surfactant [76], a similar avoidance mechanism may be involved here in the interactions between worms and the live *M. xanthus* S strain. The same compounds may also modulate interactions with the live N strain, as the mutation which knocks out A-motility may not affect production of the relevant compounds. The extent to which such compounds are repellant for the worms may be modulated by the basal prey species, as distinct metabolites may be produced in the context of different multispecies setups [77]. *M. xanthus* may produce them only during predation, or they may be repellant only compared to the more attractive compounds produced by the basal prey. If such compounds are discovered, it would be of interest to investigate whether they have specific targets or can repel a broad range of predators, to assess the importance of chemical warfare in *M. xanthus'* trophic interactions.

Our results, however, do not support the hypothesis that *M. xanthus* facultatively produces repellant compounds in response to the presence of *C. elegans*, as dead bacteria (that no longer produce any compounds) were not collectively less repellant to the worms than live bacteria (Figure 2). Alternatively, we can hypothesize that repellant compounds produced constitutively or with a different original purpose during the growth phase of *M. xanthus* remain active, at least partially, in the inoculum and deter *C. elegans* from interacting with the bacterial predator even after it is dead. For strain N in the binary choice assay, the worms that stayed on the plate were more likely to interact with the dead bacteria after 18 h than after only 1 h (Figure 2B). It is possible that after 18 h the repellant compounds had been diluted or degraded to a level that allowed *C. elegans* to interact with the dead cells. We might expect this effect to be observed more noticeably in strain N than in strain S because strain N's inability to swarm might result in a higher concentration of any secreted compounds in the vicinity of the living bacterial colony, creating a stronger contrast between the live and dead treatments of strain N than of strain S. On plates with live *M. xanthus*, we observed that the worms which entered areas with *M. xanthus* tended to aggregate around fruiting bodies. Cells undergoing development would likely decrease active production of repellant compounds in order to devote cellular resources to the developmental process. These hypotheses merit further investigation.

The impact of apex predators on mesopredators goes beyond killing effects to include indirect behavioral changes. Some predator-induced behavior changes do not require direct contact between the predator and its prey. Animal mesopredators commonly observe traces of an apex predator (e.g., scat) and modify their foraging strategies to avoid certain areas or times of day in order to reduce their own predation risk [50,78–80]. Such non-lethal effects, often called risk effects, shape not only mesopredators' behavior but also their reproduction and survival [81,82], with cascading impacts on ecosystem structure [83]. In our model system, the presence of *C. elegans* in the arena modified the predatory behavior of *M. xanthus*, even though the worms rarely interacted directly with the *M. xanthus* swarm. This effect was modulated by the basal prey identity, suggesting that prey species may exert a potential bottom-up control on the interactions between predators [84–86]. When the basal prey was *F. johnsoniae, M. xanthus'* swarming rate on the prey patch depended on the density of worms. In contrast, *M. xanthus* swarmed faster on *E. coli* in the presence of *C. elegans* irrespective of worm density (after an initial delay in the response; Figure 5). It is possible that lower attraction of the worms to *F. johnsoniae* explains the density-dependent response in *M. xanthus*: when the density of *C. elegans* is low, there are often, by chance, only very few worms in the vicinity of *M. xanthus* when it preys on *F. johnsoniae* as opposed to when it preys on *E. coli*, which is statistically less likely as the worm density increases. Such differential effect of *C. elegans* on the interaction between *M. xanthus* and the basal prey could further result from the worms carrying cells of the bacterial predator to new locations as they move around the dish. However, we would expect to see this represented in the growth pattern of *M. xanthus* on the plate by the end of the experiment, for example, as tendrils of growth emanating away from the main bacterial colony. Even after 8 days, we still saw a very distinct edge of the *M. xanthus* swarm and no visible growth outside of the main patch. We, therefore, expect the *M. xanthus* cells that could have been moved to new locations by worms to play a negligible role in the swarming rate of the *M. xanthus* patch.

The Mesopredator Release Hypothesis (MRH, e.g., [66,87]) predicts that interference interactions between apex predators and mesopredators can have profound effects on regional ecosystem structures and large-scale biomass distribution patterns. These effects have been observed in studies of animal communities [65,88]. According to the MRH, reduction in an apex predator population liberates mesopredators both from killing effects and from the need for risk-reduction behaviors. As a result, mesopredator populations increase and individuals forage more freely, which can decimate prey populations. Interference effects between top predators and mesopredators should, therefore, be considered in order to understand how ecosystems are shaped. Johnke and colleagues [25] showed that, in microbes, the combination of generalist, semi-specialist, and specialist predators can help maintain overall diversity and prevent extinction of prey species due to interference competition among the predators. However, these predators did not ne-cessarily prey on each other, and it is not known how the addition of such effects may have altered the outcome. Such studies provide valuable insight into factors maintaining diversity in microbial communities, but they are unable to address questions about more complex trophic dynamics and, in particular, the ways in which apex predators may control populations of microbial mesopredators.

In metazoans, behavioral observation often constitutes a key source of information about indirect interactions that, as previously mentioned, can alter both food web structure and dynamics, sometimes more strongly than density-mediated effects (reviewed by Werner and Peacor [89]). Our results do not yet provide a clear picture of the factors go-verning interactions between *C. elegans* and *M. xanthus*, but they offer a starting point for developing model experimental systems that allow systematic behavioral observation in nematodes and bacteria. Given the crucial role of organism behavior in the structuring of metazoan food webs, we emphasize the need for microbial food web studies to investigate behavior-mediated effects as well as direct killing effects. We suggest that *C. elegans* may not be an ideal candidate for the role of apex predator, given unclarity regarding its ability to prey on *M. xanthus*, but perhaps a protozoan or another nematode such as *Pristionchus pacificus* would more readily feed on *M. xanthus*.

In microbial communities, the overlap between ecological and evolutionary time scales has generated a number of insightful studies on food web dynamics and between-predator interactions [88,90–93]. However, most work has, to date, focused on density-mediated effects of interactions, and conceptual strategies for studying behaviors of predators of microbes remain scarce. Our synthetic community constitutes one step forward in that direction.

Supplementary Materials: The following are available online at https://www.mdpi.com/article/10.3390/microorganisms9071362/s1. Figure S1: Localization of *C. elegans* on plates containing only *M. xanthus*. Figure S2: *C. elegans* prefers prey patches not containing *M. xanthus*. Figure S3: Choice indices for prey-patch controls. Figure S4: Localization of *C. elegans* on binary choice assay plates. Table S1: Presence of prey bacteria in mixed patches.

Author Contributions: Conceptualization, G.J.V., K.A.S. and M.V.; methodology, N.M., K.A.S. and M.V.; formal analysis, N.M., K.A.S. and M.V.; investigation, N.M., K.A.S. and M.V.; writing—original draft preparation, N.M., K.A.S. and M.V.; writing—review and editing, N.M., G.J.V., K.A.S. and M.V.; visualization, K.A.S. and M.V.; supervision, K.A.S. and M.V.; project administration, K.A.S. and M.V.; funding acquisition, G.J.V. and M.V. All authors have read and agreed to the published version of the manuscript.

Funding: This research was supported in part by Swiss National Science Foundation (SNSF) grants 31003A/B_16005 to G.J.V. and an ETH Fellowship 16-2 FEL-59 to M.V.

Institutional Review Board Statement: Ethical review and approval were waived for this study because nematodes are invertebrates and are therefore not protected by EU, Swiss, or Canton of Zurich animal testing laws.

Informed Consent Statement: Not applicable.

Data Availability Statement: The data presented in this study are openly available on the Dryad repository at https://doi.org/10.5061/dryad.cvdncjt3v, accessed on 15 June 2021.

Acknowledgments: Some strains were provided by the CGC, which is funded by NIH Office of Research Infrastructure Programs (P40 OD010440). The bleaching procedure was developed with the help of Silvan Spiri and Andrea Haag from the Group of Alex Hajnal at the Institute of Molecular Life Sciences, University of Zurich. G.J.V. acknowledges Ralf Sommer and Peter Zee for stimulating earlier discussions of and P.Z. for preliminary experiments with nematode-myxobacteria-basal prey communities.

Conflicts of Interest: The authors declare no conflict of interest. The funders had no role in the design of the study; in the collection, analyses, or interpretation of data; in the writing of the manuscript; or in the decision to publish the results.

References

1. Hairston, N.G.; Smith, F.E.; Slobodkin, L.B. Community Structure, Population Control, and Competition. *Am. Nat.* **1960**, *94*, 421–425. [CrossRef]
2. Leibold, M.A.; Chase, J.M.; Shurin, J.B.; Downing, A.L. Species Turnover and the Regulation of Trophic Structure. *Annu. Rev. Ecol. Syst.* **1997**, *28*, 467–494. [CrossRef]
3. Brodie, E.D.; Brodie, E.D. Predator-Prey Arms RacesAsymmetrical selection on predators and prey may be reduced when prey are dangerous. *BioScience* **1999**, *49*, 557–568. [CrossRef]
4. Hass, C.C.; Valenzuela, D. Anti-predator benefits of group living in white-nosed coatis (*Nasua narica*). *Behav. Ecol. Sociobiol.* **2002**, *51*, 570–578. [CrossRef]
5. Matz, C.; Webb, J.S.; Schupp, P.J.; Phang, S.Y.; Penesyan, A.; Egan, S.; Steinberg, P.; Kjelleberg, S. Marine biofilm bacteria evade eukaryotic predation by targeted chemical defense. *PLoS ONE* **2008**, 3, e2744. [CrossRef] [PubMed]
6. Sinclair, A.R.E.; Mduma, S.; Brashares, J.S. Patterns of predation in a diverse predator–prey system. *Nature* **2003**, *425*, 288–290. [CrossRef] [PubMed]
7. Stevens, M.; Merilaita, S. Animal camouflage: Current issues and new perspectives. *Philos. Trans. R. Soc. B Biol. Sci.* **2009**, *364*, 423–427. [CrossRef] [PubMed]
8. Tarnita, C.E. The ecology and evolution of social behavior in microbes. *J. Exp. Biol.* **2017**, *220*, 18–24. [CrossRef] [PubMed]
9. Johnke, J.; Cohen, Y.; de Leeuw, M.; Kushmaro, A.; Jurkevitch, E.; Chatzinotas, A. Multiple micro-predators controlling bacterial communities in the environment. *Curr. Opin. Biotechnol.* **2014**, *27*, 185–190. [CrossRef]
10. Fenchel, T.; Blackburn, H.; King, G.M.; Blackburn, T.H. *Bacterial Biogeochemistry: The Ecophysiology of Mineral Cycling*; Academic Press: Cambridge, MA, USA, 2012; ISBN 978-0-12-415836-8.
11. Griffiths, B.S. Microbial-feeding nematodes and protozoa in soil: Their effecton microbial activity and nitrogen mineralization in decomposition hotspots and the rhizosphere. *Plant Soil* **1994**, *164*, 25–33. [CrossRef]
12. Kadouri, D.E.; To, K.; Shanks, R.M.Q.; Doi, Y. Predatory bacteria: A potential ally against multidrug-resistant Gram-negative pathogens. *PLoS ONE* **2013**, *8*, e63397. [CrossRef]
13. Erken, M.; Lutz, C.; McDougald, D. The rise of pathogens: Predation as a factor driving the evolution of human pathogens in the environment. *Microb. Ecol.* **2013**, *65*, 860–868. [CrossRef]
14. Davidov, Y.; Jurkevitch, E. Predation between prokaryotes and the origin of eukaryotes. *BioEssays* **2009**, *31*, 748–757. [CrossRef]
15. Forterre, P. The Common Ancestor of Archaea and Eukarya Was Not an Archaeon. *Archaea* **2013**, *2013*, e372396. [CrossRef] [PubMed]
16. Guerrero, R.; Pedros-Alio, C.; Esteve, I.; Mas, J.; Chase, D.; Margulis, L. Predatory prokaryotes: Predation and primary consumption evolved in bacteria. *Proc. Natl. Acad. Sci. USA* **1986**, *83*, 2138–2142. [CrossRef] [PubMed]
17. O'Malley, M.A. The first eukaryote cell: An unfinished history of contestation. *Stud. Hist. Philos. Biol. Biomed. Sci.* **2010**, *41*, 212–224. [CrossRef]
18. Claessen, D.; Rozen, D.E.; Kuipers, O.P.; Søgaard-Andersen, L.; van Wezel, G.P. Bacterial solutions to multicellularity: A tale of biofilms, filaments and fruiting bodies. *Nat. Rev. Microbiol.* **2014**, *12*, 115–124. [CrossRef]
19. Lyons, N.A.; Kolter, R. On the evolution of bacterial multicellularity. *Curr. Opin. Microbiol.* **2015**, *24*, 21–28. [CrossRef] [PubMed]
20. Gralka, M.; Szabo, R.; Stocker, R.; Cordero, O.X. Trophic Interactions and the Drivers of Microbial Community Assembly. *Curr. Biol.* **2020**, *30*, R1176–R1188. [CrossRef] [PubMed]
21. Kauffman, K.J.; Prakash, P.; Edwards, J.S. Advances in flux balance analysis. *Curr. Opin. Biotechnol.* **2003**, *14*, 491–496. [CrossRef] [PubMed]
22. Orth, J.D.; Thiele, I.; Palsson, B.Ø. What is flux balance analysis? *Nat. Biotechnol.* **2010**, *28*, 245–248. [CrossRef]
23. Edwards, J.S.; Covert, M.; Palsson, B. Metabolic modelling of microbes: The flux-balance approach. *Environ. Microbiol.* **2002**, *4*, 133–140. [CrossRef]
24. Ram, Y.; Dellus-Gur, E.; Bibi, M.; Karkare, K.; Obolski, U.; Feldman, M.W.; Cooper, T.F.; Berman, J.; Hadany, L. Predicting microbial growth in a mixed culture from growth curve data. *Proc. Natl. Acad. Sci. USA* **2019**, *116*, 14698–14707. [CrossRef]
25. Johnke, J.; Baron, M.; de Leeuw, M.; Kushmaro, A.; Jurkevitch, E.; Harms, H.; Chatzinotas, A. A Generalist Protist Predator Enables Coexistence in Multitrophic Predator-Prey Systems Containing a Phage and the Bacterial Predator Bdellovibrio. *Front. Ecol. Evol.* **2017**, *5*. [CrossRef]
26. Miele, V.; Guill, C.; Ramos-Jiliberto, R.; Kéfi, S. Non-trophic interactions strengthen the diversity—functioning relationship in an ecological bioenergetic network model. *PLoS Comput. Biol.* **2019**, *15*, e1007269. [CrossRef]
27. McClean, D.; Friman, V.-P.; Finn, A.; Salzberg, L.I.; Donohue, I. Coping with multiple enemies: Pairwise interactions do not predict evolutionary change in complex multitrophic communities. *Oikos* **2019**, *128*, 1588–1599. [CrossRef]

28. Johnke, J.; Boenigk, J.; Harms, H.; Chatzinotas, A. Killing the killer: Predation between protists and predatory bacteria. *FEMS Microbiol. Lett.* **2017**, *364*. [CrossRef]
29. Chen, H.; Athar, R.; Zheng, G.; Williams, H.N. Prey bacteria shape the community structure of their predators. *ISME J.* **2011**, *5*, 1314–1322. [CrossRef]
30. Chauhan, A.; Cherrier, J.; Williams, H.N. Impact of sideways and bottom-up control factors on bacterial community succession over a tidal cycle. *Proc. Natl. Acad. Sci. USA* **2009**, *106*, 4301–4306. [CrossRef]
31. Griffin, J.N.; Byrnes, J.E.K.; Cardinale, B.J. Effects of predator richness on prey suppression: A meta-analysis. *Ecology* **2013**, *94*, 2180–2187. [CrossRef]
32. Kandel, P.P.; Pasternak, Z.; van Rijn, J.; Nahum, O.; Jurkevitch, E. Abundance, diversity and seasonal dynamics of predatory bacteria in aquaculture zero discharge systems. *FEMS Microbiol. Ecol.* **2014**, *89*, 149–161. [CrossRef]
33. Pérez, J.; Moraleda-Muñoz, A.; Marcos-Torres, F.J.; Muñoz-Dorado, J. Bacterial predation: 75 years and counting! *Environ. Microbiol.* **2016**, *18*, 766–779. [CrossRef]
34. Berleman, J.E.; Scott, J.; Chumley, T.; Kirby, J.R. Predataxis behavior in *Myxococcus xanthus*. *Proc. Natl. Acad. Sci. USA* **2008**, *105*, 17127–17132. [CrossRef] [PubMed]
35. Morgan, A.D.; MacLean, R.C.; Hillesland, K.L.; Velicer, G.J. Comparative analysis of *Myxococcus* predation on soil bacteria. *Appl. Environ. Microbiol.* **2010**, *76*, 6920–6927. [CrossRef] [PubMed]
36. Nair, R.R.; Vasse, M.; Wielgoss, S.; Sun, L.; Yu, Y.-T.N.; Velicer, G.J. Bacterial predator-prey coevolution accelerates genome evolution and selects on virulence-associated prey defences. *Nat. Commun.* **2019**, *10*, 4301. [CrossRef]
37. Petters, S.; Groß, V.; Söllinger, A.; Pichler, M.; Reinhard, A.; Bengtsson, M.M.; Urich, T. The soil microbial food web revisited: Predatory myxobacteria as keystone taxa? *ISME J.* **2021**. [CrossRef] [PubMed]
38. Berleman, J.E.; Kirby, J.R. Deciphering the hunting strategy of a bacterial wolfpack. *FEMS Microbiol. Rev.* **2009**, *33*, 942–957. [CrossRef] [PubMed]
39. Livingstone, P.G.; Millard, A.D.; Swain, M.T.; Whitworth, D.E. Transcriptional changes when *Myxococcus xanthus* preys on *Escherichia coli* suggest myxobacterial predators are constitutively toxic but regulate their feeding. *Microb. Genom.* **2018**, *4*, e000152. [CrossRef]
40. Pham, V.D.; Shebelut, C.W.; Diodati, M.E.; Bull, C.T.; Singer, M. Mutations affecting predation ability of the soil bacterium *Myxococcus xanthus*. *Microbiol. Read. Engl.* **2005**, *151*, 1865–1874. [CrossRef] [PubMed]
41. Berleman, J.E.; Chumley, T.; Cheung, P.; Kirby, J.R. Rippling Is a Predatory Behavior in *Myxococcus xanthus*. *J. Bacteriol.* **2006**, *188*, 5888–5895. [CrossRef]
42. Hillesland, K.L.; Lenski, R.E.; Velicer, G.J. Ecological variables affecting predatory success in *Myxococcus xanthus*. *Microb. Ecol.* **2007**, *53*, 571–578. [CrossRef] [PubMed]
43. Mendes-Soares, H.; Velicer, G.J. Decomposing predation: Testing for parameters that correlate with predatory performance by a social bacterium. *Microb. Ecol.* **2013**, *65*, 415–423. [CrossRef]
44. Hillesland, K.L.; Velicer, G.J.; Lenski, R.E. Experimental evolution of a microbial predator's ability to find prey. *Proc. R. Soc. B Biol. Sci.* **2009**, *276*, 459–467. [CrossRef]
45. Rønn, R.; McCaig, A.E.; Griffiths, B.S.; Prosser, J.I. Impact of Protozoan Grazing on Bacterial Community Structure in Soil Microcosms. *Appl. Environ. Microbiol.* **2002**, *68*, 6094–6105. [CrossRef] [PubMed]
46. Prugh, L.R.; Stoner, C.J.; Epps, C.W.; Bean, W.T.; Ripple, W.J.; Laliberte, A.S.; Brashares, J.S. The Rise of the Mesopredator. *BioScience* **2009**, *59*, 779–791. [CrossRef]
47. Dahl, J.L.; Ulrich, C.H.; Kroft, T.L. Role of phase variation in the resistance of *Myxococcus xanthus* fruiting bodies to *Caenorhabditis elegans* predation. *J. Bacteriol.* **2011**, *193*, 5081–5089. [CrossRef]
48. Steffan, S.A.; Chikaraishi, Y.; Currie, C.R.; Horn, H.; Gaines-Day, H.R.; Pauli, J.N.; Zalapa, J.E.; Ohkouchi, N. Microbes are trophic analogs of animals. *Proc. Natl. Acad. Sci. USA* **2015**, *112*, 15119–15124. [CrossRef] [PubMed]
49. Heithaus, M.R. Habitat selection by predators and prey in communities with asymmetrical intraguild predation. *Oikos* **2001**, *92*, 542–554. [CrossRef]
50. Ritchie, E.G.; Johnson, C.N. Predator interactions, mesopredator release and biodiversity conservation. *Ecol. Lett.* **2009**, *12*, 982–998. [CrossRef]
51. Hodgkin, J.; Kaiser, D. Genetics of gliding motility in *Myxococcus xanthus* (Myxobacterales): Two gene systems control movement. *Mol. Genet. Genom. MGG* **1979**, *171*, 177–191. [CrossRef]
52. Velicer, G.J.; Raddatz, G.; Keller, H.; Deiss, S.; Lanz, C.; Dinkelacker, I.; Schuster, S.C. Comprehensive mutation identification in an evolved bacterial cooperator and its cheating ancestor. *Proc. Natl. Acad. Sci. USA* **2006**, *103*, 8107–8112. [CrossRef]
53. Velicer, G.J.; Yu, Y.N. Evolution of novel cooperative swarming in the bacterium *Myxococcus xanthus*. *Nature* **2003**, *425*, 75–78. [CrossRef] [PubMed]
54. Stiernagle, T. Maintenance of *C. elegans*. *WormBook* **2006**. [CrossRef]
55. Brenner, S. The genetics of *Caenorhabditis elegans*. *Genetics* **1974**, *77*, 71–94. [CrossRef]
56. Hagen, D.C.; Bretscher, A.P.; Kaiser, D. Synergism between morphogenetic mutants of *Myxococcus xanthus*. *Dev. Biol.* **1978**, *64*, 284–296. [CrossRef]
57. Pires-daSilva, A. *Pristionchus pacificus* protocols. *WormBook* **2013**, 1–20. [CrossRef]

58. Kauffman, A.; Parsons, L.; Stein, G.; Wills, A.; Kaletsky, R.; Murphy, C. *C. elegans* Positive Butanone Learning, Short-term, and Long-term Associative Memory Assays. *JoVE J. Vis. Exp.* **2011**, e2490. [CrossRef] [PubMed]
59. Bretscher, A.P.; Kaiser, D. Nutrition of *Myxococcus xanthus*, a fruiting myxobacterium. *J. Bacteriol.* **1978**, *133*, 763–768. [CrossRef]
60. Schindelin, J. Fiji: An open-source platform for biological-image analysis. *Nat. Methods* **2012**, *9*, 676–682. [CrossRef]
61. RStudio | Open Source & Professional Software for Data Science Teams. Available online: https://rstudio.com/ (accessed on 10 December 2020).
62. R: The R Project for Statistical Computing. Available online: https://www.r-project.org/ (accessed on 10 December 2020).
63. Lenth, R.V.; Buerkner, P.; Herve, M.; Love, J.; Riebl, H.; Singmann, H. Emmeans: Estimated Marginal Means, aka Least-Squares Means. 2020. Available online: https://cran.r-project.org/web/packages/emmeans/index.html (accessed on 10 December 2020).
64. Moore, R.S.; Kaletsky, R.; Murphy, C.T. Piwi/PRG-1 Argonaute and TGF-β Mediate Transgenerational Learned Pathogenic Avoidance. *Cell* **2019**, *177*, 1827–1841.e12. [CrossRef]
65. Brook, L.A.; Johnson, C.N.; Ritchie, E.G. Effects of predator control on behaviour of an apex predator and indirect consequences for mesopredator suppression. *J. Appl. Ecol.* **2012**, *49*, 1278–1286. [CrossRef]
66. Crooks, K.R.; Soulé, M.E. Mesopredator release and avifaunal extinctions in a fragmented system. *Nature* **1999**, *400*, 563–566. [CrossRef]
67. Wikenros, C.; Ståhlberg, S.; Sand, H. Feeding under high risk of intraguild predation: Vigilance patterns of two medium-sized generalist predators. *J. Mammal.* **2014**, *95*, 862–870. [CrossRef]
68. Félix, M.-A.; Duveau, F. Population dynamics and habitat sharing of natural populations of *Caenorhabditis elegans* and *C. briggsae*. *BMC Biol.* **2012**, *10*, 59. [CrossRef]
69. Shtonda, B.B.; Avery, L. Dietary choice behavior in *Caenorhabditis elegans*. *J. Exp. Biol.* **2006**, *209*, 89–102. [CrossRef] [PubMed]
70. Meisel, J.D.; Kim, D.H. Behavioral avoidance of pathogenic bacteria by *Caenorhabditis elegans*. *Trends Immunol.* **2014**, *35*, 465–470. [CrossRef] [PubMed]
71. Keane, R.; Berleman, J. The predatory life cycle of *Myxococcus xanthus*. *Microbiol. Read. Engl.* **2016**, *162*, 1–11. [CrossRef]
72. Berleman, J.E.; Allen, S.; Danielewicz, M.A.; Remis, J.P.; Gorur, A.; Cunha, J.; Hadi, M.Z.; Zusman, D.R.; Northen, T.R.; Witkowska, H.E.; et al. The lethal cargo of *Myxococcus xanthus* outer membrane vesicles. *Front. Microbiol.* **2014**, *5*, 474. [CrossRef]
73. Thiery, S.; Kaimer, C. The Predation Strategy of *Myxococcus xanthus*. *Front. Microbiol.* **2020**, *11*, 2. [CrossRef]
74. Findlay, B.L. The Chemical Ecology of Predatory Soil Bacteria. *ACS Chem. Biol.* **2016**, *11*, 1502–1510. [CrossRef]
75. Pradel, E.; Zhang, Y.; Pujol, N.; Matsuyama, T.; Bargmann, C.I.; Ewbank, J.J. Detection and avoidance of a natural product from the pathogenic bacterium Serratia marcescens by *Caenorhabditis elegans*. *Proc. Natl. Acad. Sci. USA* **2007**, *104*, 2295–2300. [CrossRef]
76. Islam, S.T.; Alvarez, I.V.; Saïdi, F.; Guiseppi, A.; Vinogradov, E.; Sharma, G.; Espinosa, L.; Morrone, C.; Brasseur, G.; Guillemot, J.-F.; et al. Modulation of bacterial multicellularity via spatio-specific polysaccharide secretion. *PLoS Biol.* **2020**, *18*, e3000728. [CrossRef] [PubMed]
77. Stasulli, N.M.; Shank, E.A. Profiling the metabolic signals involved in chemical communication between microbes using imaging mass spectrometry. *FEMS Microbiol. Rev.* **2016**, *40*, 807–813. [CrossRef]
78. Wikenros, C.; Jarnemo, A.; Frisén, M.; Kuijper, D.P.J.; Schmidt, K. Mesopredator behavioral response to olfactory signals of an apex predator. *J. Ethol.* **2017**, *35*, 161–168. [CrossRef]
79. Hauzy, C.; Hulot, F.D.; Gins, A.; Loreau, M. Intra- and interspecific density-dependent dispersal in an aquatic prey–predator system. *J. Anim. Ecol.* **2007**, *76*, 552–558. [CrossRef] [PubMed]
80. Shores, C.R.; Dellinger, J.A.; Newkirk, E.S.; Kachel, S.M.; Wirsing, A.J. Mesopredators change temporal activity in response to a recolonizing apex predator. *Behav. Ecol.* **2019**, *30*, 1324–1335. [CrossRef]
81. Creel, S.; Christianson, D.; Liley, S.; Winnie, J.A. Predation Risk Affects Reproductive Physiology and Demography of Elk. *Science* **2007**, *315*, 960. [CrossRef] [PubMed]
82. Creel, S.; Christianson, D. Relationships between direct predation and risk effects. *Trends Ecol. Evol.* **2008**, *23*, 194–201. [CrossRef] [PubMed]
83. Ripple, W.J.; Beschta, R.L. Wolves and the Ecology of Fear: Can Predation Risk Structure Ecosystems? *BioScience* **2004**, *54*, 755–766. [CrossRef]
84. Elmhagen, B.; Rushton, S.P. Trophic control of mesopredators in terrestrial ecosystems: Top-down or bottom-up? *Ecol. Lett.* **2007**, *10*, 197–206. [CrossRef]
85. Lynam, C.P.; Llope, M.; Möllmann, C.; Helaouët, P.; Bayliss-Brown, G.A.; Stenseth, N.C. Interaction between top-down and bottom-up control in marine food webs. *Proc. Natl. Acad. Sci. USA* **2017**, *114*, 1952–1957. [CrossRef]
86. Whalen, M.A.; Duffy, J.E.; Grace, J.B. Temporal shifts in top-down vs. bottom-up control of epiphytic algae in a seagrass ecosystem. *Ecology* **2013**, *94*, 510–520. [CrossRef] [PubMed]
87. Elmhagen, B.; Ludwig, G.; Rushton, S.P.; Helle, P.; Lindén, H. Top predators, mesopredators and their prey: Interference ecosystems along bioclimatic productivity gradients. *J. Anim. Ecol.* **2010**, *79*, 785–794. [CrossRef] [PubMed]
88. Ellner, S.P.; Becks, L. Rapid prey evolution and the dynamics of two-predator food webs. *Theor. Ecol.* **2010**, *4*, 133–152. [CrossRef]
89. Werner, E.E.; Peacor, S.D. A Review of Trait-Mediated Indirect Interactions in Ecological Communities. *Ecology* **2003**, *84*, 1083–1100. [CrossRef]

90. Hiltunen, T.; Jones, L.E.; Ellner, S.P.; Hairston, N.G. Temporal dynamics of a simple community with intraguild predation: An experimental test. *Ecology* **2013**, *94*, 773–779. [CrossRef]
91. Hiltunen, T.; Ayan, G.B.; Becks, L. Environmental fluctuations restrict eco-evolutionary dynamics in predator–prey system. *Proc. R. Soc. B Biol. Sci.* **2015**, *282*, 20150013. [CrossRef] [PubMed]
92. Kaitala, V.; Hiltunen, T.; Becks, L.; Scheuerl, T. Co-evolution as an important component explaining microbial predator-prey interaction. *J. Theor. Biol.* **2020**, *486*, 110095. [CrossRef] [PubMed]
93. Scheuerl, T.; Cairns, J.; Becks, L.; Hiltunen, T. Predator coevolution and prey trait variability determine species coexistence. *Proc. R. Soc. B Biol. Sci.* **2019**, *286*, 20190245. [CrossRef]

microorganisms

Communication

Assessment of Evolutionary Relationships for Prioritization of Myxobacteria for Natural Product Discovery

Andrew Ahearne [1,†], Hanan Albataineh [1,†], Scot E. Dowd [2] and D. Cole Stevens [1,*]

1 Department of BioMolecular Sciences, School of Pharmacy, University of Mississippi, Oxford, MS 38677, USA; aahearne@go.olemiss.edu (A.A.); haalbata@go.olemiss.edu (H.A.)
2 MR DNA, Molecular Research LP, Shallowater, TX 79363, USA; sdowd@mrdnalab.com
* Correspondence: stevens@olemiss.edu; Tel.: +1-662-915-5730
† These authors contributed equally to this work.

Abstract: Discoveries of novel myxobacteria have started to unveil the potentially vast phylogenetic diversity within the family Myxococcaceae and have brought about an updated approach to myxobacterial classification. While traditional approaches focused on morphology, 16S gene sequences, and biochemistry, modern methods including comparative genomics have provided a more thorough assessment of myxobacterial taxonomy. Herein, we utilize long-read genome sequencing for two myxobacteria previously classified as *Archangium primigenium* and *Chondrococcus macrosporus*, as well as four environmental myxobacteria newly isolated for this study. Average nucleotide identity and digital DNA–DNA hybridization scores from comparative genomics suggest previously classified as *A. primigenium* to instead be a novel member of the genus *Melittangium*, *C. macrosporus* to be a potentially novel member of the genus *Corallococcus* with high similarity to *Corallococcus exercitus*, and the four isolated myxobacteria to include another novel *Corallococcus* species, a novel *Pyxidicoccus* species, a strain of *Corallococcus exiguus*, and a potentially novel *Myxococcus* species with high similarity to *Myxococcus stipitatus*. We assess the biosynthetic potential of each sequenced myxobacterium and suggest that genus-level conservation of biosynthetic pathways support our preliminary taxonomic assignment. Altogether, we suggest that long-read genome sequencing benefits the classification of myxobacteria and improves determination of biosynthetic potential for prioritization of natural product discovery.

Keywords: myxobacteria; *Myxococcus* sp.; *Corallococcus* sp.; *Melittangium* sp.; *Archangium* sp.; biosynthetic gene clusters

Citation: Ahearne, A.; Albataineh, H.; Dowd, S.E.; Stevens, D.C. Assessment of Evolutionary Relationships for Prioritization of Myxobacteria for Natural Product Discovery. *Microorganisms* **2021**, *9*, 1376. https://doi.org/10.3390/microorganisms9071376

Academic Editors: David Whitworth and Ulrich Stingl

Received: 31 March 2021
Accepted: 21 June 2021
Published: 24 June 2021

1. Introduction

Over the last decade, 34 novel species of myxobacteria have been described including representatives from 10 newly described genera within the order *Myxococcales* (Table S1) [1–14]. Prevalent in soils and marine sediments, predatory and cellulolytic myxobacteria contribute to nutrient cycling within microbial food webs. Perhaps most-studied for their cooperative lifestyles, myxobacteria have been an excellent resource for investigations concerning developmental multicellularity and two-component signaling, swarming motilities and predatory features, and the discovery of biologically active metabolites [15–23]. Each of these areas of interest have benefited from the increased utility and accessibility of next-generation sequencing (NGS) technologies. The driving force behind the recent surge in efforts to discover novel species of myxobacteria can also be attributed to advances in sequencing technologies. Genome sequencing of myxobacteria has demonstrated that they possess large genomes replete with biosynthetic gene clusters, and myxobacteria have recently been deemed a "gifted" taxon for the production of specialized metabolites with drug-like properties [24–29]. These efforts, combined with a thorough metabolic survey of over 2000 strains within the order *Myxococcales*, concluded that the odds of novel metabolite discovery increase when exploring novel genera of myxobacteria [30]. Motivated by these observations, we sought to isolate novel myxobacteria from lesser-studied North American soils.

Recently, comparative genomic analyses have been utilized to provide efficient preliminary classification of novel myxobacteria, and we considered that such an approach would expedite prioritization of strains for future metabolic studies [3,8,11,31–37]. While traditional myxobacterial classification efforts relied on morphology, biochemistry, and the conservation of 16S gene sequences, updated methods including genome-based taxonomy have provided excellent preliminary taxonomic classification of myxobacterial isolates [38–40]. Considering that genome sequencing would also afford the biosynthetic potential of any isolated myxobacteria, we sought to employ long-read sequencing to generate high-quality draft genomes hoping to avoid fragmented, partial biosynthetic pathways. For example, of the 11 currently sequenced myxobacteria from the genus *Corallococcus*, 68% of the 621 total putative biosynthetic gene clusters (BGCs) predicted by the analysis platform AntiSMASH are positioned on a contig edge and are potentially incomplete (Table S2). In fact, the only two *Corallococcus* genomes sequenced with long-read techniques (*Corallococcus coralloides* DSM 2259^{T} and *C. coralloides* strain B035) each included 34 predicted BGCs with none located on a contig edge [41,42]. Ideally, larger contigs generated from long-read sequencing might benefit the comparative genomic analyses and provide a more complete assessment of biosynthetic potential.

In addition to four environmental isolates of putative myxobacteria included in this study, we acquired two previously characterized myxobacteria from the American Type Culture Collection (ATCC): *Archangium primigenium* ATCC 29,037 and *Chondrococcus macrosporus* ATCC 29039. Previously miscategorized as *Polyangium primigenium*, the original morphological descriptions for *A. primigenium* were remarkably apt for the strain acquired from the ATCC and cultivated in our lab, including obvious fruiting body formation and carotene-like pigmentation (Figure 1) [43,44]. The original description of *A. primigenium* fruiting bodies initially piqued our interest in the strain as members of the genus *Archangium* typically do not or very rarely form defined fruiting bodies when cultivated with standard laboratory conditions [45,46]. *Archangium* species have previously been referred to as "degenerate forms" of myxobacteria due to diminished fruiting bodies with no sporangioles or absent fruiting body formation [46]. Comparatively, little historical data is available for *C. macrosporus* ATCC 29039. The strain was deposited at the ATCC by distinguished taxonomist Professor V. B. D. Skerman and was subsequently included in a methodology study focused on isolating myxobacteria from soils [47–49]. The decision to change the genus *Chondrococcus* to instead be *Corallococcus* has been validated with many novel *Corallococcus* species being described afterwards [8,40,50]. However, we were curious to determine the status of *C. macrosporus* ATCC 29039. Considering the proposed reassignment of *Corallococcus macrosporus* DSM 14697^{T} to the genus *Myxococcus*, it was unclear if *C. macrosporus* ATCC 29,039 should also be reassigned. Both characterized using traditional approaches that heavily relied on morphology, we sought to determine how genomic comparisons might impact the current taxonomic assignments of these available myxobacteria.

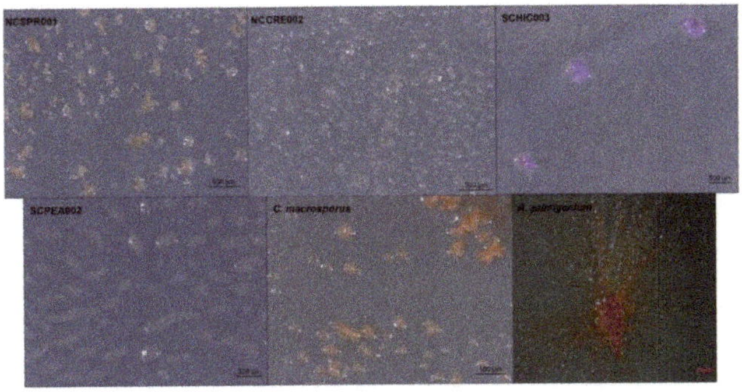

Figure 1. Myxobacterial fruiting bodies from strains NCSPR001, NCCRE002, SCHIC003, SCPEA002, and the strains *C. macrosporus* ATCC 29,039 and *A. primigenium* ATCC 29037.

2. Materials and Methods

2.1. Bacterial Strains and Growth Conditions

A. primigenium and *C. macrosporus* were procured from the ATCC as strain numbers ATCC 29037 and ATCC 29039, respectively. The remaining strains were isolated from soil as described later. All strains were cultured either on VY/2 or VY/4 agar plates (5 or 2.5 g/L baker's yeast, 1.5 g/L $CaCl_2 \cdot 2H_2O$, 0.5 mg/L vitamin B12, 15 g/L agar, pH 7.2). Swarming and fruiting bodies on agar plates were observed under a Zeiss discovery V12 stereo microscope and photographed using a Zeiss axiocam105.

2.2. Isolation of Environmental Myxobacteria

Soil samples, collected in Asheville, NC and Tryon, SC, were taken from the base of trees and dried in open air before storage. Detailed location data are provided as Supplemental Information (Table S3). Myxobacteria were isolated using a slightly modified Coli-spot method [51]. A 1 mg/mL solution of cycloheximide/nystatin was used to wet the soil sample to a paste-like consistency before inoculation onto an *Escherichia coli* baited WAT agar plate (1 g/L $CaCl_2 \cdot 2H_2O$, 15 g/L agar, 20 mM HEPES). To prepare the baiting plate, a lawn of *E. coli* was grown overnight on tryptone soya broth (TSB) with agar (1.5%), and the cells were scraped and suspended in 2 mL of sterile deionized water. Four hundred µL of the *E. coli* suspension was spread over the surface of a WAT agar plate to create a bait circle of approximately 2 inches in diameter and let dry. Once the *E. coli* was dried, a pea sized amount of soil paste was placed at the center of the bait circle. Plates were incubated at 25 °C for up to a month, and degradation of the *E. coli* was monitored over time. Visible degrading swarms were seen after a few days, and swarm edges or fruiting bodies were passaged onto VY/4 media for purification. Purification was accomplished by repeated swarm edge transfer.

2.3. Genomic DNA Isolation, Sequencing, Assembly, and Annotation

Genomic DNA for NGS was obtained from actively growing bacteria on VY/2 or VY/4 plates using NucleoBond high molecular weight DNA kit (Macherey-Nagel, Bethlehem, PA, USA). The quantity and quality of the extraction were checked by Nanodrop (Thermo Scientific NanoDrop One) and followed by Qubit quantification using Qubit®® dsDNA HS Assay Kit (ThermoFisher Scientific, Suwanee, GA, USA).

Sequencing for all samples was performed on a Pacific Biosciences single-molecule real-time (SMRT) sequencing platform at the MR DNA facility (Shallowater, TX, USA). The SMRTbell libraries for the sample were prepared using the SMRTbell Express Template Prep Kit 2.0 (Pacific Biosciences, Menlo Park, CA, USA) following the manufacturer's user guide. Following library preparation, the final concentration of each library was measured using the Qubit® dsDNA HS Assay Kit (ThermoFisher Scientific, Suwanee, GA, USA), and the average library sizes were determined using the Agilent 2100 Bioanalyzer

(Agilent Technologies, Santa Clara, CA, USA). Each library pool was then sequenced using the 10-h movie time on the PacBio Sequel (Pacific Biosciences, Menlo Park, CA, USA). De Novo Assembly of each genome was accomplished using the PacBio SMRT Analysis Hierarchical Genome Assembly Process (HGAP). Genome annotation was done using Rapid Annotation using Subsystem Technology (RAST) with further annotation requested by the NCBI Prokaryotic Genome Annotation Pipeline [52]. Sequencing data have been deposited in NCBI under the accession numbers JADWYI000000000.1, JAFIMU000000000, JAFIMS000000000, JAFIMT000000000, CP071090, and CP071091 for strains *A. primigenium*, *C. macrosporus*, NCSPR001, NCCRE002, SCPEA002, and SCHIC003, respectively.

2.4. Comparative Genomic Studies

The genome sequence data were uploaded to the Type (Strain) Genome Server (TYGS), a free bioinformatics platform available under https://tygs.dsmz.de (accessed 10 January 2021), for a whole genome-based taxonomic analysis. TYGS was used to calculate the dDDH values and construct minimum evolution trees using the Genome BLAST Distance Phylogeny approach (GBDP) [53,54]. GBDP trees were visualized using MEGA-X [55]. The average nucleotide identity (ANI) was calculated using the ANI/AAI-Matrix calculator [56,57].

2.5. BIG-SCAPE Analysis

Genome data for all myxobacteria belonging to the *Cystobacterineae* suborder were downloaded from the NCBI database. A list of all myxobacteria used in this analysis are listed in List S1. These genomes in addition to genomes of *A. primigenium*, *C. macrosporus*, and the environmental isolates were analyzed by the AntiSMASH platform (version 5 available at https://docs.antismash.secondarymetabolites.org; accessed 1 February 2021) to assess specialized metabolite gene clusters using the "relaxed" strictness setting [58,59]. A total of 1826 predicted BGCs (.gbk files) were then processed locally using the BiG-SCAPE program (version 20181005, available at https://git.wageningenur.nl/medema-group/BiG-SCAPE; accessed 1 February 2021), with the MiBIG database (version 2.0 available at https://mibig.secondarymetabolites.org; accessed 1 February 2021) as reference [60,61]. BiG-SCAPE analysis was supplemented with Pfam database version 33.1 [62]. The singleton parameter in BiG-SCAPE was selected to ensure that BGCs with distances lower than the default cutoff distance of 0.3 were included in the corresponding output data. The hybrids-off parameter was selected to prevent hybrid BGC redundancy. Generated network files separated by BiG-SCAPE class were combined for visualization using Cytoscape version 3.8.2 (http://www.cytoscape.org; accessed 1 February 2021) [63]. Annotations associated with each BGC were included in Cytoscape networks by importing curated tables generated by BiG-SCAPE.

3. Results

3.1. Comparative Genomics and Taxonomic Assessment of Archangium Primigenium, Chondrococcus Macrosporus, and Environmental Isolates

Genome sequencing provided high quality draft genomes for each of the six investigated myxobacteria, as indicated by the summary of general features in Table 1. The total genome sizes ranged from ~9.5–13 Mb, and the %GC content varied around ~69–71%. Of the six genomes, both environmental strains SCHIC003 and SCPEA002 were assembled on a single contig. Overall, the assemblies for each genome provided much lower total contig counts (1–17) than recently sequenced myxobacterial genomes [3,8]. Interestingly, a minimum evolution of phylogenetic trees generated from the whole genome sequence data clustered *A. primigenium* with *Melittangium boletus* DSM 14713^{T} and not with the three currently sequenced strains from the genus *Archangium* (Figure 2, Figures S1 and S2). Accordingly, ANI and dDDH values supported the placement of *A. primigenium* in the genus *Melittangium* (Table 2) as a novel species with both values well below the established cutoffs for classification of distinct species (<95% ANI; <70% dDDH) [31,34,35,37,64]. These data suggest *A. primigenium* is currently misclassified as a member of the genus *Archangium* and should instead be placed in the genus *Melittangium*.

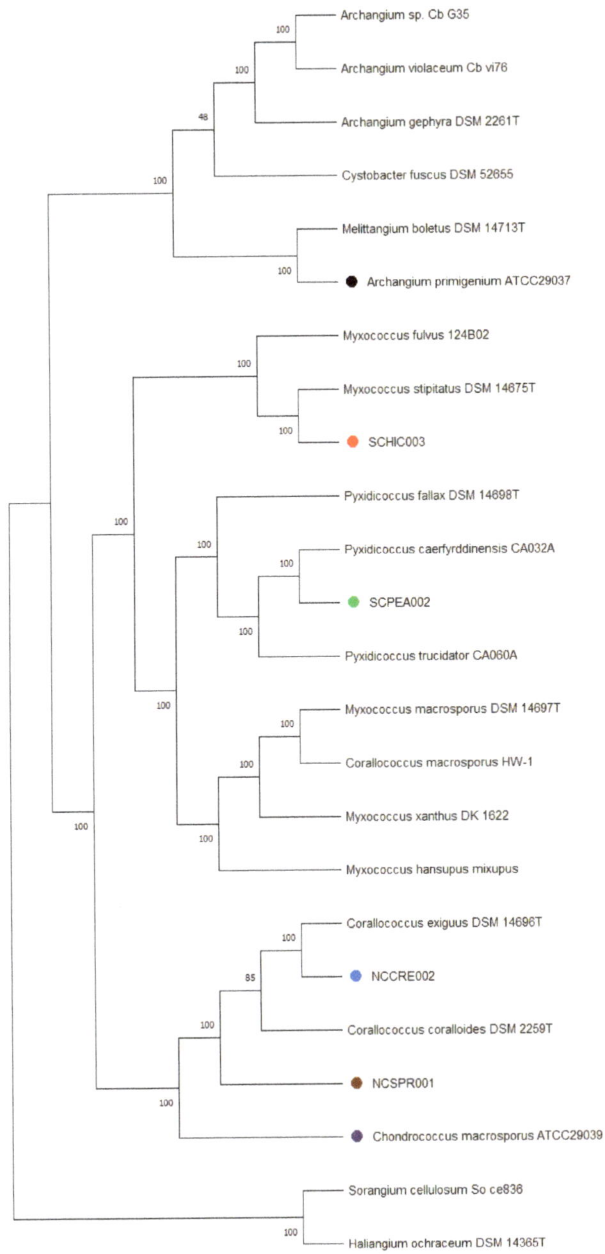

Figure 2. Minimum evolution tree from the whole genomes of different myxobacteria including the six strains under investigation in this study using the GBDP approach. The numbers in bold above branches are GBDP pseudo-bootstrap support values > 60% from 100 replications, with an average branch support of 100.0%. Branch pseudo-bootstraps less than 50% are not shown. The numbers below branches are branch lengths scaled in terms of GBDP distance formula d5. The tree was rooted at the midpoint.

Table 1. Genome properties and general features of myxobacteria under investigation in this study.

Species	Size (bp)	CDS	GC%	N50	L50	Contigs	Coverage
A. primigenium	9,491,554	7873	70.7%	9,468,833	1	3	441x
C. macrosporus	9,811,739	7977	70.4%	1,094,727	2	17	300x
NCSPR001	9,785,177	8033	70.1%	9,343,940	1	3	312x
NCCRE002	10,538,407	8589	69.7%	3,024,381	2	8	479x
SCPEA002	13,211,253	10,588	69.6%	N/A	1	1	144x
SCHIC003	10,367,529	8339	68.6%	N/A	1	1	301x

Table 2. 16S rRNA identity, ANI, and dDDH values for pairwise comparisons between *A. primigenium* with the most similar fully sequenced myxobacteria.

Species	16s rRNA	dDDH	ANI
M. boletus DSM 14713^T	98.89%	29.5	86.1%
C. fuscus DSM 52655	98.7%	24.5	83.29%
A. gephyra DSM 2261^T	97.72%	23.2	81.41%
S. aurantiaca DW43-1	96.06%	20	78.9%
M. macrosporus DSM 14697^T	96.63%	19.8	78.85%

The calculated ANI and dDDH values for the sequenced *C. marcosporus* strain acquired from the ATCC support the original assignment to the genus *Chondrococcus*, now *Corallococcus* [31,50]. As opposed to the recently reclassified *Myxococcus macrosporus* DSM 14697^T, previously *Corallococcus macrosporus*, the minimum evolution phylogenetic tree suggested *C. macrosporus* ATCC 29039 to be a member or the genus *Corallococcus* most similar to *Corallococcus exercitus* DSM 108849^T (Figure 2, Figures S1 and S3) [50]. The isolated strains NCCRE002 and NCSPR001 were also determined to be members of the genus *Corallococcus* (Figure 2, Figures S1 and S3). Comparative genome analyses implied that strain NCCRE002 is an isolate of *Corallococcus exiguus* DSM 14696^T. However, the ANI and GBDP trees suggested that strain NCSPR001 is a novel member of the genus *Corallococcus* most similar to *Corallococcus coralloides* DSM 2259^T (Table 3).

The isolated SCHIC003 and SCPEA002 strains were initially determined to be members of the genus *Myxococcus*. However, inclusion of sequenced representatives from the genus *Pyxidicoccus* (considered to be synonymous with *Myxococcus*) [3] in our comparative analysis grouped strain SCPEA002 within the *Pyxidicoccus* clade (Figure 2, Figures S1 and S4). Most similar to *Pyxidicoccus caerfyrddinensis* CA032A^T, dDDH and ANI analysis suggested the SCPEA002 strain to be a novel member of the genus *Pyxidicoccus* (Table 4). Similarly, comparative genome analysis determined that strain SCHIC003 is likely be a novel member of the genus *Myxococcus*, albeit highly similar to *Myxococcus stipitatus* DSM 14675^T with ANI and dDDH values just below the cutoffs for species differentiation [31,37,64] (Table 4 and Figure 2, Figures S1 and S3).

Table 3. Differentiation chart comparing *C. macrosporus* ATCC 29039, NCCRE002, and NCSPR001 draft genome data with sequenced members of the genus *Corallococcus*. The top half uses total genome comparison methods (ANI and dDDH) while the bottom half uses 16S rRNA sequence for pairwise comparison. Orange shading represents species that would be designated as the same using the designated method. Blue shading represents unique species using the designated method, <98.65% 16S identity%, or < 95%/70% for ANI/dDDH.

	NCCRE002	NCSPR001	*Chondrococcus macrosporus*	*Corallococcus interemptor* T	*Corallococcus terminator* T	*Corallococcus sicarius* T	*Corallococcus praedator* T	*Corallococcus macrosporus* HW1	*Corallococcus llansteffanensis* T	*Corallococcus exiguus* T	*Corallococcus exercitus* T	*Corallococcus coralloides* T	*Corallococcus carmarthensis* T	*Corallococcus aberystwythensis* T	*Corallococcus* Z5C101001	*Corallococcus* ZKHCc1_1396	*Corallococcus* CA053C
NCCRE002	100%	dDDH: 51 ANI: 94	41 91	44 92	29 86	29 86	30 86	21 81	30 87	66 96	43 91	54 94	43 92	43 91	34 88	29 86	30 86
NCSPR001	99.8	100%	dDDH: 41 ANI: 91	45 92	29 86	30 87	30 87	21 81	30 87	51 93	43 92	54 94	43 91	43 91	34 88	29 87	30 87
Chondrococcus macrosporus	99.15	99.22	100%	dDDH: 42 ANI: 91	30 86	30 87	30 87	21 81	31 87	41 91	51 94	42 91	43 92	44 92	34 89	30 87	31 87
Corallococcus interemptor T	99.8	99.87	99.35	100%	dDDH: 29 ANI: 86	30 87	30 87	21 81	30 87	44 92	42 91	46 92	42 91	42 91	34 88	30 87	30 87
Corallococcus terminator T	99.03	98.96	99.09	98.96	100%	dDDH: 35 ANI: 89	49 93	21 81	35 89	29 86	30 87	29 86	30 87	30 87	31 87	42 91	34 89
Corallococcus sicarius T	98.89	98.83	98.98	98.83	99.61	100%	dDDH: 35 ANI: 89	21 81	50 93	30 86	31 87	30 87	31 87	31 87	31 87	43 89	50 93
Corallococcus praedator T	99.03	98.96	99.09	98.96	100	99.61	100%	dDDH: 21 ANI: 81	36 89	30 86	31 87	30 87	31 87	31 87	32 87	43 92	35 89
Corallococcus macrosporus HW1	97.73	97.66	98.37	97.79	97.72	97.72	97.72	100%	dDDH: 22 ANI: 81	21 81	21 81	21 81	21 81	21 81	21 81	21 81	21 81
Corallococcus llansteffanensis T	98.83	98.76	98.89	98.76	99.54	99.93	99.54	97.73	100%	dDDH: 30 ANI: 87	32 88	31 87	32 87	31 87	33 88	36 89	54 94
Corallococcus exiguus T	99.93	99.87	99.22	99.87	99.09	98.96	99.09	97.79	98.89	100%	dDDH: 43 ANI: 91	54 94	44 92	43 91	34 88	30 86	30 86
Corallococcus exercitus T	99.02	99.09	99.87	99.22	99.22	99.09	99.22	98.37	99.02	99.09	100%	dDDH: 44 ANI: 92	48 93	47 93	36 89	31 87	32 87
Corallococcus coralloides T	99.67	99.61	99.09	99.74	98.83	98.7	98.83	97.66	98.63	99.74	98.96	100%	dDDH: 44 ANI: 92	44 92	34 88	30 87	30 87
Corallococcus carmarthensis T	99.22	99.28	99.93	99.28	99.15	99.02	99.15	98.31	98.96	99.28	99.8	99.02	100%	dDDH: 48 ANI: 93	36 89	31 87	31 87
Corallococcus aberystwythensis T	99.35	99.15	99.8	99.15	99.15	99.02	99.15	98.31	98.96	99.26	99.67	99.02	99.87	100%	dDDH: 35 ANI: 89	31 87	31 87
Corallococcus Z5C101001	98.76	98.83	99.48	98.83	99.22	99.22	99.22	98.11	99.15	98.83	99.61	98.57	99.54	99.41	100%	dDDH: 32 ANI: 88	32 88
Corallococcus ZKHCc1_1396	98.7	98.89	98.89	98.76	99.67	99.35	99.67	97.4	99.28	98.76	99.02	98.5	98.96	98.83	99.09	100%	dDDH: 35 ANI: 88
Corallococcus CA053C	98.57	98.5	98.7	98.5	99.54	99.67	99.54	97.59	99.61	98.63	98.83	98.37	98.76	98.76	99.22	99.28	100%

Table 4. Differentiation chart comparing SCPEA002 and SCHIC003 draft genome data with sequenced members of the genera *Myxococcus* and *Pyxidicoccus*. The top half uses total genome comparison methods (ANI and dDDH) while the bottom half uses 16S rRNA sequence for pairwise comparison. Orange shading represents species that would be designated as the same using the designated method. Blue shading represents unique species using the designated method, <98.65% 16S identity%, or < 95%/70% for ANI/dDDH.

	SCPEA002	SCHIC003	*M. fulvus* 124B02	*P. fallax* T	*M. stipitatus* T	*M. hansupus*	*M. eversor* T	*M. llanfair* T	*M. vastator* T	*M. virescens* T	*P. trucidator* T	*P. caerfyrd-dinensis* T	*M. xanthus* DK1622	*M. macrosporus* T
SCPEA002	100%	dDDH: 22 ANI: 82	23 82	28 85	22 82	23 83	23 82	28 82	25 83	24 83	29 86	34 88	24 83	24 83
SCHIC003	99.15%	100%	dDDH: 26 ANI: 84	23 82	49 93	22 81	27 85	27 85	23 82	22 82	22 82	22 82	22 81	22 82
***M. fulvus* 124B02**	99.61%	99.41%	100%	dDDH: 23 ANI: 82	26 84	22 82	28 85	28 85	23 82	23 82	23 82	23 82	22 82	23 82
***P. fallax* T**	99.54%	98.96%	99.41%	100%	dDDH: 23 ANI: 82	24 83	23 82	23 82	25 84	25 83	30 86	29 86	25 83	25 84
***M. stipitatus* T**	99.15%	100%	99.41%	98.96%	100%	dDDH: 22 ANI: 81	27 85	27 85	23 82	22 82	22 82	22 82	29 81	22 82
M. hansupus	98.89%	98.44%	98.89%	98.57%	98.44%	100%	dDDH: 22 ANI: 82	23 82	32 88	32 87	24 83	24 83	31 87	32 88
***M. eversor* T**	98.83%	98.24%	98.83%	98.50%	98.24%	99.15%	100%	dDDH: 41 ANI: 91	23 82	22 82	24 82	23 82	22 82	23 82
***M. llanfair* T**	98.76%	98.31%	98.89%	98.44%	98.31%	99.09%	99.93%	100%	dDDH: 23 ANI: 82	23 82	24 83	23 82	23 82	23 82
***M. vastator* T**	98.70%	98.24%	98.70%	98.37%	98.24%	99.41%	98.96%	98.89%	100%	dDDH: 52 ANI: 94	25 84	25 84	52 94	41 91
***M. virescens* T**	98.63%	98.14%	98.63%	98.31%	98.18%	99.35%	98.89%	98.83%	99.67%	100%	dDDH: 25 ANI: 83	24 83	73 97	40 90
***P. trucidator* T**	99.09%	98.37%	98.70%	99.02%	98.37%	98.89%	98.83%	98.76%	98.70%	98.63%	100%	dDDH: 33 ANI: 88	24 83	25 84
***P. caerfyrddinensis* T**	99.48%	98.76%	99.09%	99.41%	98.76%	99.15%	98.96%	98.89%	98.96%	98.96%	99.61%	100%	dDDH: 24 ANI: 83	25 83
***M. xanthus* DK1622**	98.57%	98.11%	98.57%	98.24%	98.11%	99.28%	98.83%	98.76%	99.74%	99.93%	98.57%	98.83%	100%	dDDH: 40 ANI: 90
***M. macrosporus* T**	98.89%	98.44%	98.89%	98.57%	98.44%	99.48%	99.15%	99.09%	99.67%	99.61%	98.89%	99.15%	99.54%	100%

3.2. Biosynthetic Potential and Genus Level Correlations

Analysis of our draft genomes using the biosynthetic pathway prediction platform AntiSMASH revealed a range of 29–42 total predicted BGCs with *C. macrosporus* including the highest total of BGCs. However, the draft genome for *C. macrosporus* also included the highest total of four partial BGCs positioned on the edges of contigs. No BGCs occurring on contig edges were observed from *A. primigenium*, NCSPR001, or SCPEA002. All of the sequenced strains included highly similar (≥75% similarity score) biosynthetic pathways for the signaling terpene geosmin [65,66], the signaling lipids VEPE/AEPE/TG-1 [67,68], and carotenoids [69–72] (Figure 3). Excluding SCHIC003, each genome included a BGC highly homologous to the pathway associated with the myxobacterial siderophore myxochelin [73,74]. Pathways somewhat similar (similarity scores of 66%) to the myxoprincomide-c506 BGC were observed in every genome except the *A. primigenium* genome [75]. Clusters with ≥75% similarity to pathways from *M. stipitatus* DSM 14675^T associated with the metabolites rhizopodin [76,77] and phenalamide A2 [78] were observed in the SCHIC003 draft genome as well as clusters also present in the *M. stipitatus* DSM 14675^T genome deposited in the AntiSMASH database [79], including the dkxanthene [80], fulvuthiacene [81], and violacein [82–84] BGCs (Figure 4). Considering previously characterized BGCs from each genus associated with the six investigated myxobacteria, the corallopyronin BGC from *C. coralloides* B035 [85,86] was absent from all three of the putative *Corallococcus* strains, the melithiazol BGC from *Melittangium lichenicola* Me I46 [87] was not present in *A. primigenium*, and neither the disciformycin/gulmirecin BGC [88,89] or the pyxidicycline BGC [90] from *Pyxidicoccus fallax* were present in SCPEA002.

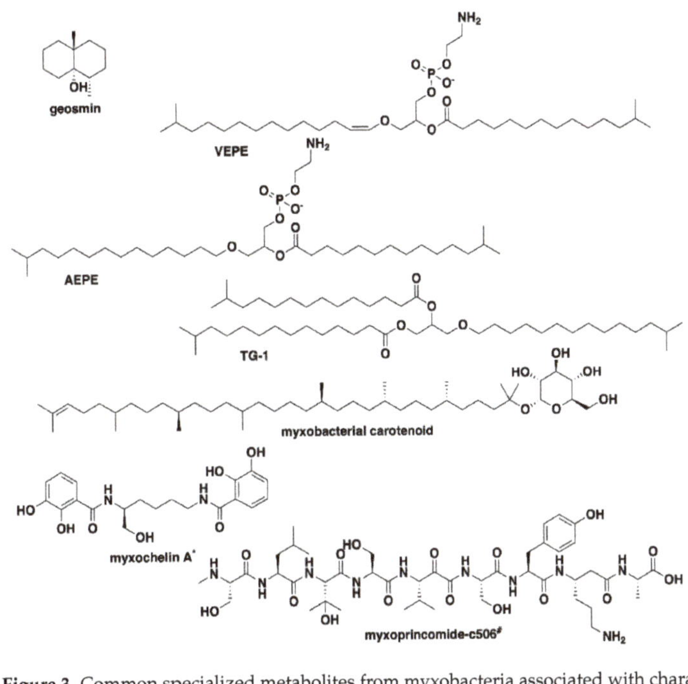

Figure 3. Common specialized metabolites from myxobacteria associated with characterized BGCs present in the six investigated strains of myxobacteria. * Myxochelin BGC not present in SCHIC003 genome data. # Myxoprincomide BGC not present in *A. primigenium* genome data.

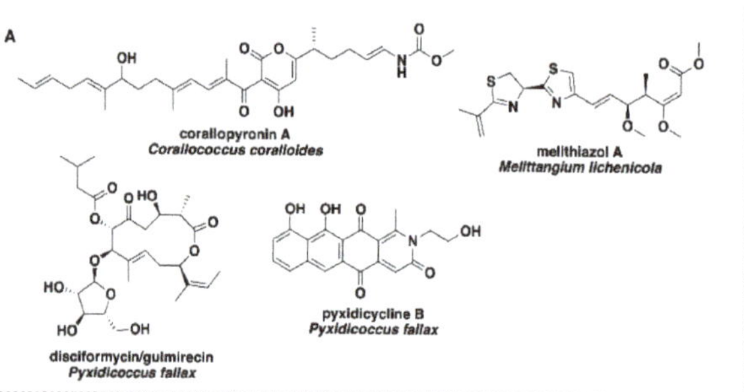

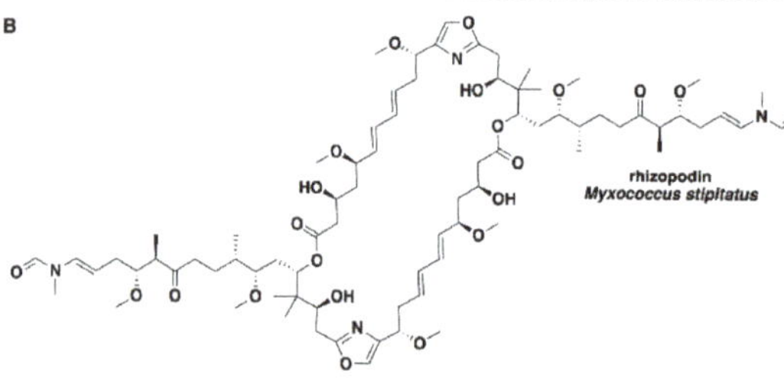

protein	SCHIC003 accession #	*M. stipitatus* accession #	description	% identity
RizA	WP_206716295.1	WP_015350862.1	ACP S-malonyltransferase	97.89
RizB	WP_206716296.1	WP_169558665.1	hybrid NRPS-transAT PKS	89.13
RizC (partial)*	WP_206716297.1	WP_169558665.1	transAT PKS	91.76
RizD	WP_206716298.1	WP_015350860.1	hybrid NRPS-transAT PKS	88.97
RizE	WP_206716299.1	WP_015350859.1	transAT PKS	86.55
RizF	WP_206716300.1	WP_015350858.1	ACP S-malonyltransferase	93.81
CCA89331.1	WP_206716301.1	WP_015350857.1	acyltransferase domain-containing protein	96.57

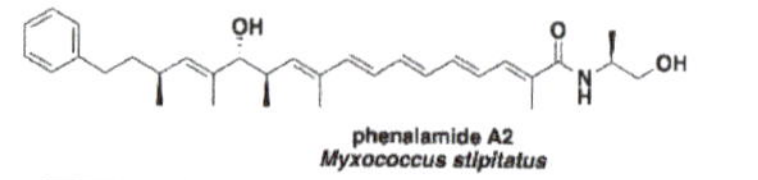

protein	SCHIC003 accession #	*M. stipitatus* accession #	description	% identity
PhaA	WP_206717160.1	WP_015349878.1	NRPS	93.83
PhaB	WP_206717159.1	WP_015349879.1	type I PKS	90.71
PhaC	WP_206717158.1	WP_015349880.1	type I PKS	92.02
PhaD	WP_206717157.1	WP_015349881.1	type I PKS	91.98
PhaE	WP_206717156.1	WP_015349882.1	type I PKS	89.50
PhaF	WP_206717155.1	WP_015349883.1	type I PKS	88.93
PhaG	WP_206717154.1	WP_015349884.1	type I PKS	89.72
PhaH	WP_206717153.1	WP_015349885.1	2-oxo acid dehydrogenase subunit E2	86.82
PhaI	WP_206717152.1	WP_015349886.1	alpha-ketoacid dehydrogenase subunit beta	97.54
PhaJ	WP_015349887.1	WP_015349887.1	pyruvate dehydrogenase	90.98

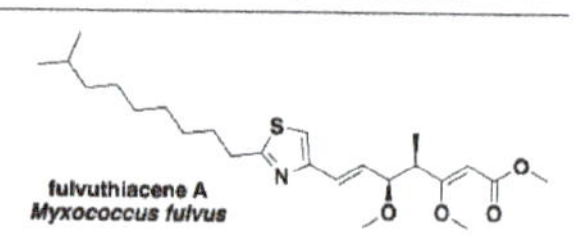

protein	SCHIC003 accession #	*M. fulvus* accession #	description	% identity
FtaA	WP_206716604.1	AXM42948.1	type I PKS	84.84
FtaB	WP_206716605.1	AXM42949.1	hybrid PKS/NRPS	89.79
FtaC	WP_206719706.1	AXM42950.1	type I PKS	90.33
FtaD	WP_206716606.1	AXM42951.1	type I PKS	95.58
FtaE	WP_206716607.1	AXM42952.1	NRPS	91.17
FtaF	WP_206716608.1	AXM42953.1	alphahydroxyglycine hydrolase-like protein	94.72
FtaG	WP_206716609.1	AXM42954.1	methyltransferase	92.20
FtaH	WP_206716610.1	AXM42955.1	amide hydrolase	85.76
FtaI	WP_206716603.1	AXM42947.1	adenosylhomocysteinase	97.69
FtaJ	WP_206716602.1	AXM42946.1	bifunctional dehydrogenase/cyclohydrolase	93.27
FtaK	WP_206716601.1	AXM42945.1	hypothetical protein	84.27
FtaL	WP_206716600.1	AXM42944.1	thioesterase	97.02
FtaM	WP_206716599.1	AXM42943.1	enoyl-CoA hydratase/isomerase	95.03
FtaN	WP_206716598.1	AXM42942.1	methyltransferase	95.03

Figure 4. (**A**) Specialized metabolites produced by members of the genera *Corallococcus, Melittangium,* and *Pyxidicoccus* with no associated BGCs observed in any of the six investigated myxobacterial strains. (**B**) Comparisons of the rhizopodin, phenalamide A2, and fulvuthiacene BGCs from SCHIC003 genome data and the characterized pathways from *M. stipitatus* and *M. fulvus.* All SCHIC003 gene products, excluding RizC, had coverages ≥99% with the indicated homolog. * RizC located on a contig edge and is incomplete in SCHIC003 genome data.

Utilizing the BiG-SCAPE platform to render BGC sequence similarity networks, we sought to determine the extent of homology between BGCs from our six sequenced myxobacteria and BGCs from all currently sequenced members of the suborder *Cystobacterineae* [91]. The resulting sequence similarity network included 1080 BGCs connected by 3046 edges (not including self-looped nodes/singletons) and depicted genus-level homologies across all BGCs from the newly sequenced myxobacteria corroborating our suggested taxonomic assignments (Figure 5 and Table 5). For example, BGCs from the three newly sequenced samples *C. macrosporus*, NCSPR001, and NCCRE002 were almost exclusively clustered with BGCs from members of the genus *Corallococcus*, and BGCs from SCHIC003 and SCPEA002 samples clustered with the genera *Myxococcus* and *Pyxidicoccus* (Figure 5). However, SCPEA002 BGCs do not cluster as frequently with *Pyxidicoccus* BGCs as they do *Myxococcus* BGCs, and the majority (76.5%) were not clustered with any BGC within the network (Table 5). This is likely due to the highly fragmented nature of available *Pyxidicoccus* genomes resulting in many incomplete or partial BGCs. Therefore, few *Pyxidicoccus* pathways appear in the similarity network, and the percentage of unique pathways associated with SCPEA002 is likely overestimated. Regardless, the limited number of SCPEA002 BGCs clustered with BGCs from *Myxococcus/Pyxidicoccus* genomes indicates a potential to discover novel metabolites despite placement in the highly scrutinized clade. The only clustered groups with numerous edges formed between BGCs from the genera

Myxococcus and *Corallococcus* included characterized biosynthetic pathways for ubiquitous signaling lipids VEPE/AEPE/TG-1, carotenoids, and the siderophore myxochelin as well as two uncharacterized BGCs predicted to produce ribosomally synthesized and post-translationally modified peptides (RiPPs).

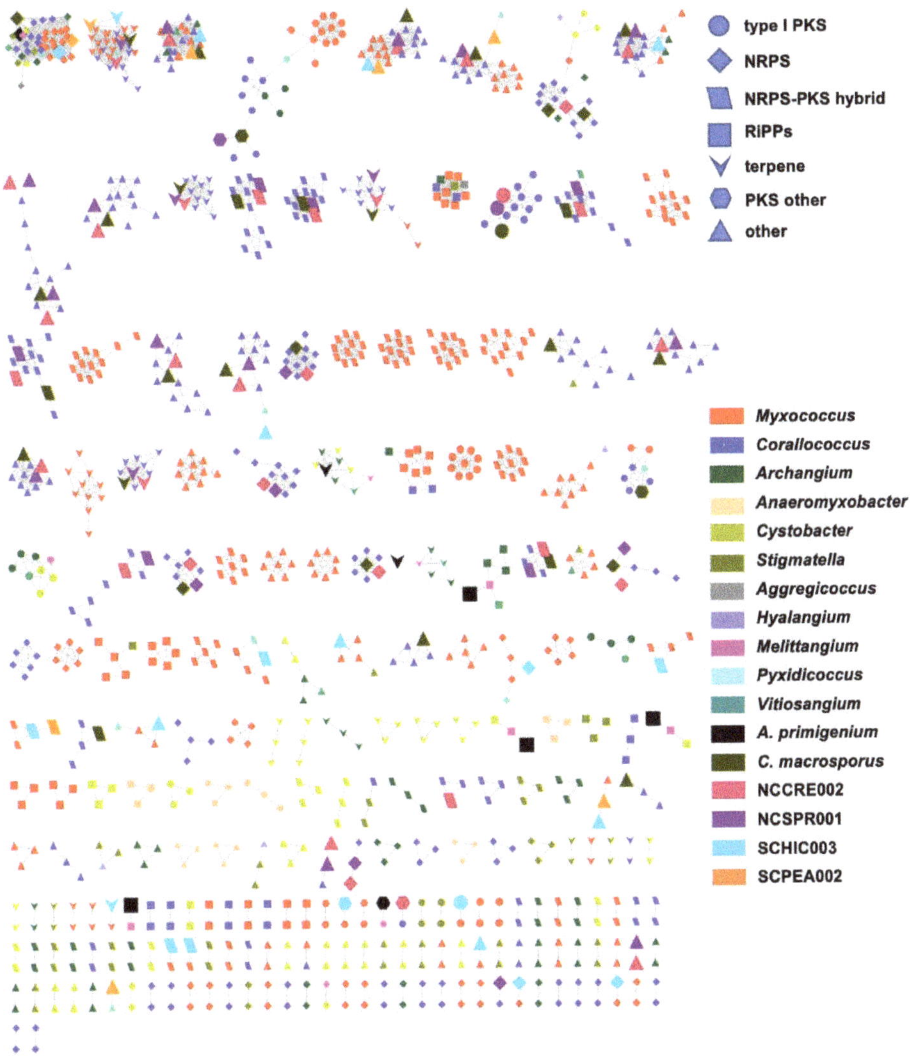

Figure 5. BiG-SCAPE BGC sequence similarity networks (c = 0.3) as visualized with Cytoscape. The network is generated from *A. primigenium*, *C. macrosporus*, NCCRE002, NCSPR001, SCHIC003, SCPEA002, and all myxobacteria belonging to the Cystobacterineae suborder with genomes deposited in NCBI. Each node represents one BGC identified by AntiSMASH 5.0, where the colors and shapes of the nodes represent different genera and AntiSMASH-predicted classes, respectively. Nodes representing BGCs from newly sequenced myxobacteria included in this study are enlarged. BGCs included as singletons in the original BiG-SCAPE analysis removed.

Table 5. Overview of BiG-SCAPE BGC sequence similarity networks of the six strains under investigation in this study.

Myxobacteria	# of Total BGCs	# and % of Singletons	# of Edges Formed with other BGCs	# of BGCs with 1 or 2 Edges	# of BGCs with 3 or More Edges
A. primigenium ATCC 29037	32	24 (75%)	21	6	2
C. macrosporus ATCC 29039 *	42	9 (21.4%)	228	4	29
NCSPR001	32	1 (3.1%)	248	7	24
NCCRE002 *	36	3 (16.7%)	231	7	26
SCPEA002	34	26 (76.5%)	62	4	4
SCHIC003	29	8 (27.6%)	85	13	8

* genomes with fragmented biosynthetic pathways, likely resulting in fewer clustered pathways than truly exist.

Interestingly, a total of 23 *A. primigenium* BGCs (out of 32 BGCs) appear as singletons in the network with no homology to any of the included BGCs from *Cystobacterineae*. In fact, aside from the VEPE/AEPE/TG-1 cluster and a terpene cluster that included members of the genera *Archangium* and *Cystobacter*, all remaining BGCs from *A. primigenium* had connecting edges to BGCs from *Melittangium boletus* DSM 14713^T. Out of 21 edges formed by *A. primigenium* in the network, four edges were formed with four species of *Corallococcus* (a total of 11 *Corallococcus* species in the network), four edges were formed with all species of *Cystobacter* (three species in the network), six edges were formed with all species of *Archangium* (three species in the network), and seven edges were formed with the only *Melittangium* species in the network, *M. boletus* DSM 14713^T. Overall, these data corroborate our preliminary taxonomic assignments and suggest that the prioritization of *A. primigenium* for subsequent discovery efforts is most likely to yield novel metabolites.

4. Discussion

As novel myxobacteria continue to be isolated and explored for natural product discovery, efficient approaches for approximate taxonomic placement will assist the prioritization of lesser studied genera. Utilizing long-read genome sequencing and comparative genomic analyses, we determine preliminary taxonomic placement for four myxobacteria isolated from North American soils and two myxobacteria deposited at the ATCC. This approach indicated that previously classified *A. primigenium* ATCC 29037 is instead a novel member of the genus *Melittangium*, and that three of our four environmental isolates included potentially novel members of the genera *Corallococcus*, *Myxococcus*, and *Pyxidicoccus*. Previously classified *Chondorococcus macrosporus* ATCC 29039 was also determined to be a potentially novel member of the genus *Corallococcus*, with high similarity to *C. exercitus* DSM 108849^T and phylogenetically distinct from *M. macrosporus* DSM 14697^T previously assigned to the genus *Corallococcus*. Subsequent bioinformatic analysis of biosynthetic pathways included in the newly sequenced genomes corroborated our preliminary taxonomic placements for each sample. Ultimately, this process identified *A. primigenium* to be a member of the lesser studied genus *Melittangium* and indicated that it should be prioritized for continued natural product discovery efforts. Of the environmental isolates, BGCs from SCPEA002 were determined to include the least amount of overlap with BGCs from other *Myxococcus/Pyxidicoccus* species. While environmental isolates SCHIC002 and NCSPR001 were also identified as novel members of the genera *Myxococcus* and *Corallococcus*, respectively, the apparent overlap in BGCs from thoroughly explored myxobacteria determined from sequence similarity network analysis suggests a limited potential for discovery of novel specialized metabolites. Overall, comparative genomic techniques including the assessment of biosynthetic potential enabled a phylogenetic approximation and suggested prioritization of *A. primigenium* for natural product discovery efforts from a sample set of six newly sequenced myxobacteria.

Supplementary Materials: The following are available online at https://www.mdpi.com/article/10.3390/microorganisms9071376/s1. Supplemental Table S1: Novel myxobacteria described in literature spanning 2011-2021; Table S2: Comparison of biosynthetic gene clusters (BGCs) located on contig edges from previously sequenced members of the genus *Corallococcus*, Table S3: Locations of soil samples used for environmental strains isolation, List S1: A list of all myxobacteria and their accession numbers used in BiG-SCAPE; Figure S1: Minimum evolution tree from the 16S rRNA of the 6 strains under investigation in this study and all myxobacteria Type strains deposited in DSMZ; Figure S2: Minimum evolution tree from the whole genomes of *A. primigenium* and different members of the family *Archangiaceae* using the GBDP approach; Figure S3: Minimum evolution tree from the whole genomes of *C. macrosporus* ATCC 29039, NCCRE002, NCSPR001 and different members of genus *Corallococcus* using the GBDP approach; Figure S4: Minimum evolution tree from the whole genomes of SCHIC003, SCPEA002, and different members of *Myxococcus* and *Pyxidicoccus* using the GBDP approach.

Author Contributions: Conceptualization, supervision, and administration D.C.S.; formal analysis and data curation A.A., H.A., S.E.D., and D.C.S.; methodology, validation, and writing A.A., H.A., and D.C.S. All authors have read and agreed to the published version of the manuscript.

Funding: This research was supported by funds from the National Institute of Allergy and Infectious Diseases (1 R15 AI137996-01A1) and the National Institute of General Medical Sciences (1 P20 GM130460-01A1).

Institutional Review Board Statement: Not applicable.

Informed Consent Statement: Not Applicable.

Data Availability Statement: Sequencing data have been deposited in NCBI under the accession numbers JADWYI000000000.1, JAFIMU000000000, JAFIMS000000000, JAFIMT000000000, CP071090, and CP071091 for strains *A. primigenium*, *C. macrosporus*, NCSPR001, NCCRE002, SCPEA002, and SCHIC003, respectively.

Acknowledgments: The authors would like to acknowledge the University of Mississippi School of Pharmacy for startup support.

Conflicts of Interest: The authors declare no conflict of interest.

References

1. Awal, R.P.; Garcia, R.; Gemperlein, K.; Wink, J.; Kunwar, B.; Parajuli, N.; Muller, R. Vitiosangium cumulatum gen. nov., sp. nov. and Vitiosangium subalbum sp. nov., soil myxobacteria, and emended descriptions of the genera Archangium and Angiococcus, and of the family Cystobacteraceae. *Int. J. Syst. Evol. Microbiol.* **2017**, *67*, 1422–1430. [CrossRef]
2. Awal, R.P.; Garcia, R.; Muller, R. Racemicystis crocea gen. nov., sp. nov., a soil myxobacterium in the family Polyangiaceae. *Int. J. Syst. Evol. Microbiol.* **2016**, *66*, 2389–2395. [CrossRef]
3. Chambers, J.; Sparks, N.; Sydney, N.; Livingstone, P.G.; Cookson, A.R.; Whitworth, D.E. Comparative Genomics and Pan-Genomics of the Myxococcaceae, including a Description of Five Novel Species: Myxococcus eversor sp. nov., Myxococcus llanfairpwllgwyngyllgogerychwyrndrobwllllantysiliogogogochensis sp. nov., Myxococcus vastator sp. nov., Pyxidicoccus caerfyrddinensis sp. nov., and Pyxidicoccus trucidator sp. nov. *Genome Biol. Evol.* **2020**, *12*, 2289–2302. [CrossRef]
4. Garcia, R.; Gemperlein, K.; Muller, R. Minicystis rosea gen. nov., sp. nov., a polyunsaturated fatty acid-rich and steroid-producing soil myxobacterium. *Int. J. Syst. Evol. Microbiol.* **2014**, *64*, 3733–3742. [CrossRef]
5. Garcia, R.; Muller, R. Simulacricoccus ruber gen. nov., sp. nov., a microaerotolerant, non-fruiting, myxospore-forming soil myxobacterium and emended description of the family Myxococcaceae. *Int. J. Syst. Evol. Microbiol.* **2018**, *68*, 3101–3110. [CrossRef]
6. Garcia, R.; Stadler, M.; Gemperlein, K.; Muller, R. Aetherobacter fasciculatus gen. nov., sp. nov. and Aetherobacter rufus sp. nov., novel myxobacteria with promising biotechnological applications. *Int. J. Syst. Evol. Microbiol.* **2016**, *66*, 928–938. [CrossRef] [PubMed]
7. Iizuka, T.; Jojima, Y.; Hayakawa, A.; Fujii, T.; Yamanaka, S.; Fudou, R. Pseudenhygromyxa salsuginis gen. nov., sp. nov., a myxobacterium isolated from an estuarine marsh. *Int. J. Syst. Evol. Microbiol.* **2013**, *63*, 1360–1369. [CrossRef]
8. Livingstone, P.G.; Ingleby, O.; Girdwood, S.; Cookson, A.R.; Morphew, R.M.; Whitworth, D.E. Predatory Organisms with Untapped Biosynthetic Potential: Descriptions of Novel Corallococcus Species C. aberystwythensis sp. nov., C. carmarthensis sp. nov., C. exercitus sp. nov., C. interemptor sp. nov., C. llansteffanensis sp. nov., C. praedator sp. nov., C. sicarius sp. nov., and C. terminator sp. nov. *Appl. Environ. Microbiol.* **2020**, *86*. [CrossRef]
9. Mohr, K.I.; Garcia, R.O.; Gerth, K.; Irschik, H.; Muller, R. Sandaracinus amylolyticus gen. nov., sp. nov., a starch-degrading soil myxobacterium, and description of Sandaracinaceae fam. nov. *Int. J. Syst. Evol. Microbiol.* **2012**, *62*, 1191–1198. [CrossRef] [PubMed]

10. Mohr, K.I.; Moradi, A.; Glaeser, S.P.; Kampfer, P.; Gemperlein, K.; Nubel, U.; Schumann, P.; Muller, R.; Wink, J. Nannocystis konarekensis sp. nov., a novel myxobacterium from an Iranian desert. *Int. J. Syst. Evol. Microbiol.* **2018**, *68*, 721–729. [CrossRef]
11. Mohr, K.I.; Wolf, C.; Nubel, U.; Szafranska, A.K.; Steglich, M.; Hennessen, F.; Gemperlein, K.; Kampfer, P.; Martin, K.; Muller, R.; et al. A polyphasic approach leads to seven new species of the cellulose-decomposing genus Sorangium, Sorangium ambruticinum sp. nov., Sorangium arenae sp. nov., Sorangium bulgaricum sp. nov., Sorangium dawidii sp. nov., Sorangium kenyense sp. nov., Sorangium orientale sp. nov. and Sorangium reichenbachii sp. nov. *Int. J. Syst. Evol. Microbiol.* **2018**, *68*, 3576–3586. [CrossRef] [PubMed]
12. Moradi, A.; Ebrahimipour, G.H.; Mohr, K.I.; Kampfer, P.; Glaeser, S.P.; Hennessen, F.; Gemperlein, K.; Awal, R.P.; Wolf, C.; Muller, R.; et al. Racemicystis persica sp. nov., a myxobacterium from soil. *Int. J. Syst. Evol. Microbiol.* **2017**, *67*, 472–478. [CrossRef]
13. Sood, S.; Awal, R.P.; Wink, J.; Mohr, K.I.; Rohde, M.; Stadler, M.; Kampfer, P.; Glaeser, S.P.; Schumann, P.; Garcia, R.; et al. Aggregicoccus edonensis gen. nov., sp. nov., an unusually aggregating myxobacterium isolated from a soil sample. *Int. J. Syst. Evol. Microbiol.* **2015**, *65*, 745–753. [CrossRef] [PubMed]
14. Yamamoto, E.; Muramatsu, H.; Nagai, K. Vulgatibacter incomptus gen. nov., sp. nov. and Labilithrix luteola gen. nov., sp. nov., two myxobacteria isolated from soil in Yakushima Island, and the description of Vulgatibacteraceae fam. nov., Labilitrichaceae fam. nov. and Anaeromyxobacteraceae fam. nov. *Int. J. Syst. Evol. Microbiol.* **2014**, *64*, 3360–3368. [CrossRef]
15. Bader, C.D.; Panter, F.; Muller, R. In depth natural product discovery–Myxobacterial strains that provided multiple secondary metabolites. *Biotechnol. Adv.* **2020**, *39*, 107480. [CrossRef] [PubMed]
16. Bretl, D.J.; Kirby, J.R. Molecular Mechanisms of Signaling in Myxococcus xanthus Development. *J. Mol. Biol.* **2016**, *428*, 3805–3830. [CrossRef] [PubMed]
17. Mercier, R.; Mignot, T. Regulations governing the multicellular lifestyle of Myxococcus xanthus. *Curr. Opin. Microbiol.* **2016**, *34*, 104–110. [CrossRef]
18. Mohr, K.I. Diversity of Myxobacteria-We Only See the Tip of the Iceberg. *Microorganisms* **2018**, *6*, 84. [CrossRef]
19. Pathak, D.T.; Wei, X.; Wall, D. Myxobacterial tools for social interactions. *Res. Microbiol.* **2012**, *163*, 579–591. [CrossRef]
20. Petters, S.; Gross, V.; Sollinger, A.; Pichler, M.; Reinhard, A.; Bengtsson, M.M.; Urich, T. The soil microbial food web revisited: Predatory myxobacteria as keystone taxa? *ISME J.* **2021**. [CrossRef]
21. Sah, G.P.; Wall, D. Kin recognition and outer membrane exchange (OME) in myxobacteria. *Curr. Opin. Microbiol.* **2020**, *56*, 81–88. [CrossRef]
22. Thiery, S.; Kaimer, C. The Predation Strategy of Myxococcus xanthus. *Front. Microbiol.* **2020**, *11*, 2. [CrossRef]
23. Whitworth, D.E. Genome-wide analysis of myxobacterial two-component systems: Genome relatedness and evolutionary changes. *BMC Genom.* **2015**, *16*, 780. [CrossRef]
24. Baltz, R.H. Molecular beacons to identify gifted microbes for genome mining. *J. Antibiot.* **2017**, *70*, 639–646. [CrossRef] [PubMed]
25. Baltz, R.H. Gifted microbes for genome mining and natural product discovery. *J. Ind. Microbiol. Biotechnol.* **2017**, *44*, 573–588. [CrossRef] [PubMed]
26. Herrmann, J.; Fayad, A.A.; Muller, R. Natural products from myxobacteria: Novel metabolites and bioactivities. *Nat. Prod. Rep.* **2017**, *34*, 135–160. [CrossRef] [PubMed]
27. Landwehr, W.; Wolf, C.; Wink, J. Actinobacteria and Myxobacteria-Two of the Most Important Bacterial Resources for Novel Antibiotics. *Curr. Top. Microbiol. Immunol.* **2016**, *398*, 273–302. [CrossRef] [PubMed]
28. Weissman, K.J.; Muller, R. Myxobacterial secondary metabolites: Bioactivities and modes-of-action. *Nat. Prod. Rep.* **2010**, *27*, 1276–1295. [CrossRef]
29. Wenzel, S.C.; Muller, R. Myxobacteria–'microbial factories' for the production of bioactive secondary metabolites. *Mol. Biosyst.* **2009**, *5*, 567–574. [CrossRef] [PubMed]
30. Hoffmann, T.; Krug, D.; Bozkurt, N.; Duddela, S.; Jansen, R.; Garcia, R.; Gerth, K.; Steinmetz, H.; Muller, R. Correlating chemical diversity with taxonomic distance for discovery of natural products in myxobacteria. *Nat. Commun.* **2018**, *9*, 803. [CrossRef]
31. Chun, J.; Oren, A.; Ventosa, A.; Christensen, H.; Arahal, D.R.; da Costa, M.S.; Rooney, A.P.; Yi, H.; Xu, X.W.; De Meyer, S.; et al. Proposed minimal standards for the use of genome data for the taxonomy of prokaryotes. *Int. J. Syst. Evol. Microbiol.* **2018**, *68*, 461–466. [CrossRef] [PubMed]
32. Chun, J.; Rainey, F.A. Integrating genomics into the taxonomy and systematics of the Bacteria and Archaea. *Int. J. Syst. Evol. Microbiol.* **2014**, *64*, 316–324. [CrossRef] [PubMed]
33. Garcia, R.; Gerth, K.; Stadler, M.; Dogma, I.J., Jr.; Muller, R. Expanded phylogeny of myxobacteria and evidence for cultivation of the 'unculturables'. *Mol. Phylogenet. Evol.* **2010**, *57*, 878–887. [CrossRef]
34. Goris, J.; Konstantinidis, K.T.; Klappenbach, J.A.; Coenye, T.; Vandamme, P.; Tiedje, J.M. DNA-DNA hybridization values and their relationship to whole-genome sequence similarities. *Int. J. Syst. Evol. Microbiol.* **2007**, *57*, 81–91. [CrossRef]
35. Kim, M.; Oh, H.S.; Park, S.C.; Chun, J. Towards a taxonomic coherence between average nucleotide identity and 16S rRNA gene sequence similarity for species demarcation of prokaryotes. *Int. J. Syst. Evol. Microbiol.* **2014**, *64*, 346–351. [CrossRef] [PubMed]
36. Livingstone, P.G.; Morphew, R.M.; Whitworth, D.E. Genome Sequencing and Pan-Genome Analysis of 23 Corallococcus spp. Strains Reveal Unexpected Diversity, With Particular Plasticity of Predatory Gene Sets. *Front. Microbiol.* **2018**, *9*, 3187. [CrossRef]
37. Sangal, V.; Goodfellow, M.; Jones, A.L.; Schwalbe, E.C.; Blom, J.; Hoskisson, P.A.; Sutcliffe, I.C. Next-generation systematics: An innovative approach to resolve the structure of complex prokaryotic taxa. *Sci. Rep.* **2016**, *6*, 38392. [CrossRef]

38. Shimkets, L.; Woese, C.R. A phylogenetic analysis of the myxobacteria: Basis for their classification. *Proc. Natl. Acad. Sci. USA* **1992**, *89*, 9459–9463. [CrossRef] [PubMed]
39. Sproer, C.; Reichenbach, H.; Stackebrandt, E. The correlation between morphological and phylogenetic classification of myxobacteria. *Int. J. Syst. Evol. Microbiol.* **1999**, *49 Pt 3*, 1255–1262. [CrossRef]
40. Stackebrandt, E.; Pauker, O.; Steiner, U.; Schumann, P.; Straubler, B.; Heibei, S.; Lang, E. Taxonomic characterization of members of the genus Corallococcus: Molecular divergence versus phenotypic coherency. *Syst. Appl. Microbiol.* **2007**, *30*, 109–118. [CrossRef]
41. Bouhired, S.; Rupp, O.; Blom, J.; Schaberle, T.F.; Schiefer, A.; Kehraus, S.; Pfarr, K.; Goesmann, A.; Hoerauf, A.; Konig, G. Complete Genome Sequence of the Corallopyronin A-Producing Myxobacterium Corallococcus coralloides B035. *Microbiol. Resour. Announc.* **2019**, *8*, e00050-19. [CrossRef] [PubMed]
42. Huntley, S.; Zhang, Y.; Treuner-Lange, A.; Kneip, S.; Sensen, C.W.; Sogaard-Andersen, L. Complete genome sequence of the fruiting myxobacterium Corallococcus coralloides DSM 2259. *J. Bacteriol.* **2012**, *194*, 3012–3013. [CrossRef]
43. McDonald, J.C. Studies on the genus Archangium (Myxobacterales). II. The effect of temperature and carbohydrates on some physiological processes. *Mycologia* **1967**, *59*, 1059–1068. [CrossRef] [PubMed]
44. McDonald, J.C. Studies on the genus Archangium (Myxobacterales) I. Morphology. *Mycologia* **1965**, *57*, 737–747. [CrossRef]
45. Shimkets, L.J.; Dworkin, M.; Reichenbach, H. The Myxobacteria. In *The Prokaryotes*; Dworkin, M., Falkow, S., Rosenberg, E., Schleifer, K.H., Stackebrandt, E., Eds.; Springer: New York, NY, USA, 2006. [CrossRef]
46. McCurdy, H.D. Studies on the Taxonomy of the Myxobacterales. *Int. J. Syst. Bacteriol.* **1971**, *21*, 50–54. [CrossRef]
47. Starr, M.P.; Skerman, V.B. Bacterial diversity: The natural history of selected morphologically unusual bacteria. *Annu. Rev. Microbiol.* **1965**, *19*, 407–454. [CrossRef] [PubMed]
48. Sly, L.I. Taxonomic Note: V. B. D. Skerman (1921–1993), a Reforming Force in Bacterial Systematics and Nomenclature. *Int. J. Syst. Evol. Microbiol.* **1995**, *45*, 412–413. [CrossRef]
49. Karwowski, J.P.; Sunga, G.N.; Kadam, S.; McAlpine, J.B. A method for the selective isolation of Myxococcus directly from soil. *J. Ind. Microbiol.* **1996**, *16*, 230–236. [CrossRef]
50. Lang, E.; Stackebrandt, E. Emended descriptions of the genera Myxococcus and Corallococcus, typification of the species Myxococcus stipitatus and Myxococcus macrosporus and a proposal that they be represented by neotype strains. Request for an Opinion. *Int. J. Syst. Evol. Microbiol.* **2009**, *59*, 2122–2128. [CrossRef]
51. Dawid, W. Biology and global distribution of myxobacteria in soils. *FEMS Microbiol. Rev.* **2000**, *24*, 403–427. [CrossRef]
52. Aziz, R.K.; Bartels, D.; Best, A.A.; DeJongh, M.; Disz, T.; Edwards, R.A.; Formsma, K.; Gerdes, S.; Glass, E.M.; Kubal, M.; et al. The RAST Server: Rapid annotations using subsystems technology. *BMC Genom.* **2008**, *9*, 75. [CrossRef]
53. Camacho, C.; Coulouris, G.; Avagyan, V.; Ma, N.; Papadopoulos, J.; Bealer, K.; Madden, T.L. BLAST+: Architecture and applications. *BMC Bioinform.* **2009**, *10*, 421. [CrossRef]
54. Meier-Kolthoff, J.P.; Goker, M. TYGS is an automated high-throughput platform for state-of-the-art genome-based taxonomy. *Nat. Commun.* **2019**, *10*, 2182. [CrossRef] [PubMed]
55. Kumar, S.; Stecher, G.; Li, M.; Knyaz, C.; Tamura, K. MEGA X: Molecular Evolutionary Genetics Analysis across Computing Platforms. *Mol. Biol. Evol.* **2018**, *35*, 1547–1549. [CrossRef] [PubMed]
56. Lee, I.; Ouk Kim, Y.; Park, S.C.; Chun, J. OrthoANI: An improved algorithm and software for calculating average nucleotide identity. *Int. J. Syst. Evol. Microbiol.* **2016**, *66*, 1100–1103. [CrossRef] [PubMed]
57. Rodriguez-R, L.M.; Konstantinidis, K.T. The enveomics collection: A toolbox for specialized analyses of microbial genomes and metagenomes. *PeerJ* **2016**, *4*, e1900v1.
58. Medema, M.H.; Blin, K.; Cimermancic, P.; de Jager, V.; Zakrzewski, P.; Fischbach, M.A.; Weber, T.; Takano, E.; Breitling, R. antiSMASH: Rapid identification, annotation and analysis of secondary metabolite biosynthesis gene clusters in bacterial and fungal genome sequences. *Nucleic Acids Res.* **2011**, *39*, W339–W346. [CrossRef]
59. Blin, K.; Shaw, S.; Steinke, K.; Villebro, R.; Ziemert, N.; Lee, S.Y.; Medema, M.H.; Weber, T. antiSMASH 5.0: Updates to the secondary metabolite genome mining pipeline. *Nucleic Acids Res.* **2019**, *47*, W81–W87. [CrossRef] [PubMed]
60. Kautsar, S.A.; Blin, K.; Shaw, S.; Navarro-Munoz, J.C.; Terlouw, B.R.; van der Hooft, J.J.J.; van Santen, J.A.; Tracanna, V.; Suarez Duran, H.G.; Pascal Andreu, V.; et al. MIBiG 2.0: A repository for biosynthetic gene clusters of known function. *Nucleic Acids Res.* **2020**, *48*, D454–D458. [CrossRef]
61. Navarro-Munoz, J.C.; Selem-Mojica, N.; Mullowney, M.W.; Kautsar, S.A.; Tryon, J.H.; Parkinson, E.I.; De Los Santos, E.L.C.; Yeong, M.; Cruz-Morales, P.; Abubucker, S.; et al. A computational framework to explore large-scale biosynthetic diversity. *Nat. Chem. Biol.* **2020**, *16*, 60–68. [CrossRef]
62. Mistry, J.; Chuguransky, S.; Williams, L.; Qureshi, M.; Salazar, G.A.; Sonnhammer, E.L.L.; Tosatto, S.C.E.; Paladin, L.; Raj, S.; Richardson, L.J.; et al. Pfam: The protein families database in 2021. *Nucleic Acids Res.* **2021**, *49*, D412–D419. [CrossRef] [PubMed]
63. Shannon, P.; Markiel, A.; Ozier, O.; Baliga, N.S.; Wang, J.T.; Ramage, D.; Amin, N.; Schwikowski, B.; Ideker, T. Cytoscape: A software environment for integrated models of biomolecular interaction networks. *Genome Res.* **2003**, *13*, 2498–2504. [CrossRef] [PubMed]
64. Richter, M.; Rossello-Mora, R. Shifting the genomic gold standard for the prokaryotic species definition. *Proc. Natl. Acad. Sci. USA* **2009**, *106*, 19126–19131. [CrossRef] [PubMed]
65. Bentley, R.; Meganathan, R. Geosmin and methylisoborneol biosynthesis in streptomycetes. Evidence for an isoprenoid pathway and its absence in non-differentiating isolates. *FEBS Lett.* **1981**, *125*, 220–222. [CrossRef]

66. Dickschat, J.S.; Bode, H.B.; Mahmud, T.; Muller, R.; Schulz, S. A novel type of geosmin biosynthesis in myxobacteria. *J. Org. Chem.* **2005**, *70*, 5174–5182. [CrossRef]
67. Bhat, S.; Ahrendt, T.; Dauth, C.; Bode, H.B.; Shimkets, L.J. Two lipid signals guide fruiting body development of Myxococcus xanthus. *mBio* **2014**, *5*, e00939-13. [CrossRef]
68. Lorenzen, W.; Bozhuyuk, K.A.; Cortina, N.S.; Bode, H.B. A comprehensive insight into the lipid composition of Myxococcus xanthus by UPLC-ESI-MS. *J. Lipid Res.* **2014**, *55*, 2620–2633. [CrossRef]
69. Botella, J.A.; Murillo, F.J.; Ruiz-Vazquez, R. A cluster of structural and regulatory genes for light-induced carotenogenesis in Myxococcus xanthus. *Eur. J. Biochem.* **1995**, *233*, 238–248. [CrossRef]
70. Cervantes, M.; Murillo, F.J. Role for vitamin B(12) in light induction of gene expression in the bacterium Myxococcus xanthus. *J. Bacteriol.* **2002**, *184*, 2215–2224. [CrossRef]
71. Lopez-Rubio, J.J.; Elias-Arnanz, M.; Padmanabhan, S.; Murillo, F.J. A repressor-antirepressor pair links two loci controlling light-induced carotenogenesis in Myxococcus xanthus. *J. Biol. Chem.* **2002**, *277*, 7262–7270. [CrossRef]
72. Perez-Marin, M.C.; Padmanabhan, S.; Polanco, M.C.; Murillo, F.J.; Elias-Arnanz, M. Vitamin B12 partners the CarH repressor to downregulate a photoinducible promoter in Myxococcus xanthus. *Mol. Microbiol.* **2008**, *67*, 804–819. [CrossRef] [PubMed]
73. Gaitatzis, N.; Kunze, B.; Muller, R. In vitro reconstitution of the myxochelin biosynthetic machinery of Stigmatella aurantiaca Sg a15: Biochemical characterization of a reductive release mechanism from nonribosomal peptide synthetases. *Proc. Natl. Acad. Sci. USA* **2001**, *98*, 11136–11141. [CrossRef]
74. Li, Y.; Weissman, K.J.; Muller, R. Myxochelin biosynthesis: Direct evidence for two- and four-electron reduction of a carrier protein-bound thioester. *J. Am. Chem. Soc.* **2008**, *130*, 7554–7555. [CrossRef]
75. Cortina, N.S.; Krug, D.; Plaza, A.; Revermann, O.; Muller, R. Myxoprincomide: A natural product from Myxococcus xanthus discovered by comprehensive analysis of the secondary metabolome. *Angew. Chem. Int. Ed.* **2012**, *51*, 811–816. [CrossRef] [PubMed]
76. Sasse, F.; Steinmetz, H.; Hofle, G.; Reichenbach, H. Rhizopodin, a new compound from Myxococcus stipitatus (myxobacteria) causes formation of rhizopodia-like structures in animal cell cultures. Production, isolation, physico-chemical and biological properties. *J. Antibiot.* **1993**, *46*, 741–748. [CrossRef] [PubMed]
77. Pistorius, D.; Muller, R. Discovery of the rhizopodin biosynthetic gene cluster in Stigmatella aurantiaca Sg a15 by genome mining. *Chembiochem* **2012**, *13*, 416–426. [CrossRef]
78. Park, S.; Hyun, H.; Lee, J.S.; Cho, K. Identification of the Phenalamide Biosynthetic Gene Cluster in Myxococcus stipitatus DSM 14675. *J. Microbiol. Biotechnol.* **2016**, *26*, 1636–1642. [CrossRef]
79. Blin, K.; Shaw, S.; Kautsar, S.A.; Medema, M.H.; Weber, T. The antiSMASH database version 3: Increased taxonomic coverage and new query features for modular enzymes. *Nucleic Acids Res.* **2021**, *49*, D639–D643. [CrossRef]
80. Meiser, P.; Weissman, K.J.; Bode, H.B.; Krug, D.; Dickschat, J.S.; Sandmann, A.; Muller, R. DKxanthene biosynthesis–understanding the basis for diversity-oriented synthesis in myxobacterial secondary metabolism. *Chem. Biol.* **2008**, *15*, 771–781. [CrossRef] [PubMed]
81. Panter, F.; Krug, D.; Muller, R. Novel Methoxymethacrylate Natural Products Uncovered by Statistics-Based Mining of the Myxococcus fulvus Secondary Metabolome. *ACS Chem. Biol.* **2019**, *14*, 88–98. [CrossRef] [PubMed]
82. Momen, A.Z.; Hoshino, T. Biosynthesis of violacein: Intact incorporation of the tryptophan molecule on the oxindole side, with intramolecular rearrangement of the indole ring on the 5-hydroxyindole side. *Biosci. Biotechnol. Biochem.* **2000**, *64*, 539–549. [CrossRef]
83. Brady, S.F.; Chao, C.J.; Handelsman, J.; Clardy, J. Cloning and heterologous expression of a natural product biosynthetic gene cluster from eDNA. *Org. Lett.* **2001**, *3*, 1981–1984. [CrossRef] [PubMed]
84. Hoshino, T. Violacein and related tryptophan metabolites produced by Chromobacterium violaceum: Biosynthetic mechanism and pathway for construction of violacein core. *Appl. Microbiol. Biotechnol.* **2011**, *91*, 1463–1475. [CrossRef]
85. Erol, O.; Schaberle, T.F.; Schmitz, A.; Rachid, S.; Gurgui, C.; El Omari, M.; Lohr, F.; Kehraus, S.; Piel, J.; Muller, R.; et al. Biosynthesis of the myxobacterial antibiotic corallopyronin A. *Chembiochem* **2010**, *11*, 1253–1265. [CrossRef] [PubMed]
86. Pogorevc, D.; Panter, F.; Schillinger, C.; Jansen, R.; Wenzel, S.C.; Muller, R. Production optimization and biosynthesis revision of corallopyronin A, a potent anti-filarial antibiotic. *Metab. Eng.* **2019**, *55*, 201–211. [CrossRef]
87. Weinig, S.; Hecht, H.J.; Mahmud, T.; Muller, R. Melithiazol biosynthesis: Further insights into myxobacterial PKS/NRPS systems and evidence for a new subclass of methyl transferases. *Chem. Biol.* **2003**, *10*, 939–952. [CrossRef]
88. Schieferdecker, S.; Konig, S.; Weigel, C.; Dahse, H.M.; Werz, O.; Nett, M. Structure and biosynthetic assembly of gulmirecins, macrolide antibiotics from the predatory bacterium Pyxidicoccus fallax. *Chemistry* **2014**, *20*, 15933–15940. [CrossRef] [PubMed]
89. Surup, F.; Viehrig, K.; Mohr, K.I.; Herrmann, J.; Jansen, R.; Muller, R. Disciformycins A and B: 12-membered macrolide glycoside antibiotics from the myxobacterium Pyxidicoccus fallax active against multiresistant staphylococci. *Angew. Chem. Int. Ed.* **2014**, *53*, 13588–13591. [CrossRef]
90. Panter, F.; Krug, D.; Baumann, S.; Muller, R. Self-resistance guided genome mining uncovers new topoisomerase inhibitors from myxobacteria. *Chem. Sci.* **2018**, *9*, 4898–4908. [CrossRef] [PubMed]
91. Gregory, K.; Salvador, L.A.; Akbar, S.; Adaikpoh, B.I.; Stevens, D.C. Survey of Biosynthetic Gene Clusters from Sequenced Myxobacteria Reveals Unexplored Biosynthetic Potential. *Microorganisms* **2019**, *7*, 781. [CrossRef] [PubMed]

microorganisms

Review

Myxobacterial Genomics and Post-Genomics: A Review of Genome Biology, Genome Sequences and Related 'Omics Studies

David E. Whitworth *, Natashia Sydney and Emily J. Radford

Institute of Biological, Environmental and Rural Sciences, Aberystwyth University, Aberystwyth SY23 3DD, UK; nas39@aber.ac.uk (N.S.); emr33@aber.ac.uk (E.J.R.)
* Correspondence: dew@aber.ac.uk

Abstract: Myxobacteria are fascinating and complex microbes. They prey upon other members of the soil microbiome by secreting antimicrobial proteins and metabolites, and will undergo multicellular development if starved. The genome sequence of the model myxobacterium *Myxococcus xanthus* DK1622 was published in 2006 and 15 years later, 163 myxobacterial genome sequences have now been made public. This explosion in genomic data has enabled comparative genomics analyses to be performed across the taxon, providing important insights into myxobacterial gene conservation and evolution. The availability of myxobacterial genome sequences has allowed system-wide functional genomic investigations into entire classes of genes. It has also enabled post-genomic technologies to be applied to myxobacteria, including transcriptome analyses (microarrays and RNA-seq), proteome studies (gel-based and gel-free), investigations into protein–DNA interactions (ChIP-seq) and metabolism. Here, we review myxobacterial genome sequencing, and summarise the insights into myxobacterial biology that have emerged as a result. We also outline the application of functional genomics and post-genomic approaches in myxobacterial research, highlighting important findings to emerge from seminal studies. The review also provides a comprehensive guide to the genomic datasets available in mid-2021 for myxobacteria (including 24 genomes that we have sequenced and which are described here for the first time).

Keywords: comparative genomics; functional genomics; genome evolution; genome organisation; pan-genome; proteomics; taxonomy; transcriptomics

Citation: Whitworth, D.E.; Sydney, N.; Radford, E.J. Myxobacterial Genomics and Post-Genomics: A Review of Genome Biology, Genome Sequences and Related 'Omics Studies. *Microorganisms* **2021**, *9*, 2143. https://doi.org/10.3390/microorganisms9102143

Academic Editor: Renato Fani

Received: 22 September 2021
Accepted: 12 October 2021
Published: 13 October 2021

1. Myxobacterial Genomics

Myxobacteria are ubiquitous soil-dwelling bacteria which have attracted considerable research interest due to their complex behaviours, ecological importance and production of potentially useful bio-active compounds [1,2]. They have rod-shaped cells, typically 0.5 microns in width by 3–8 microns in length, which can move backwards and forwards over a surface, in the direction of the long axis of the cell. Myxobacteria are found abundantly and ubiquitously in soils, but have also been found in virtually all other environments where they have been looked for. As apex microbial predators, they are able to kill and feed on the other microbes they encounter. Prey killing is achieved by the secretion of antimicrobial metabolites and digestive enzymes into the public commons, which cause prey cell lysis [3–5]. When prey and nutrients are scarce, myxobacteria instead undergo cooperative multicellular development. A population of 100,000 cells orchestrate their movements to aggregate together into fruiting bodies. Some species make fruiting bodies which are simple raised mounds, while other myxobacteria make complex tree-shaped structures. Within fruiting bodies, a subset of cells differentiate into metabolically dormant and stress-resistant myxospores [6], which germinate when nutrients/prey become available again.

The most thoroughly studied myxobacterium is *Myxococcus xanthus* DK1622, which has been investigated since the 1970s [7]. Over the following decades, a genetic toolbox was developed for DK1622 and other *M. xanthus* strains (including transposons, reporter genes, targeted mutagenesis systems, transducing phage, and gene insertions), which enabled elucidation of the genetic basis of fruiting body development and other myxobacterial behaviours [8,9].

Prior to genome sequencing, it was already known that myxobacteria typically had large genomes [10,11], and that their genomes had a high (~70%) %GC content [12,13] (the percentage of G–C base pairs in the genome). Later, cloning and sequencing of individual genes confirmed the GC-rich nature of myxobacterial genomes, and also observed reading frame bias in protein-coding genes [14,15]. To maintain an overall %GC in protein-coding sequences (CDSs), the three bases of the triplet codon have distinctive %GC biases, and the pattern of those biases can be helpful when identifying protein-coding features in genome sequences [16].

1.1. The Genome Sequence of M. xanthus DK1622

The first myxobacterial genome sequence to be made publicly available was that of *Anaeromyxobacter dehalogenans* 2CP-C [17]. However, the model myxobacterium DK1622 was the first myxobacterium to have its genome sequence described in the literature [18]. Sequencing of *M. xanthus* DK1622 was initially performed by Monsanto, producing a draft genome known as the M1 genome, which remains unpublished. Using the M1 genome, complete sequencing was then undertaken by The Institute of Genome Research (TIGR), and annotation of the resulting genome sequence was undertaken as a project involving the entire myxobacteria research community.

When sequenced, *M. xanthus* DK1622's exact genome size was revealed to be 9.14 Mbp, comprised of a single circular chromosome with a %GC content of 68.9. The large size of the genome compared with other bacteria was proposed to have been due in part to lineage-specific duplications generating paralagous genes [18]. Later, Huntley et al. [19] investigated the possibility that the large size of fruiting myxobacterial genomes was due to whole genome duplication, but found no evidence for that scenario, instead reinforcing the importance of gene duplications. Such duplications seemed to have disproportionately affected genes encoding signalling proteins (Ser/Thr protein kinases, two-component signal transduction systems), transcriptional regulators, chemosensory and motility proteins. It was suggested that the increase in regulatory potential from duplications of genes encoding signalling proteins and regulators, allowed the evolution of the complex regulatory network necessary for multicellular fruiting body formation [18].

Using BLAST, Goldman et al. [18] also found that one-third of the CDSs of *M. xanthus* DK1622 were most similar to genes from distantly related organisms (outside class Deltaproteobacteria, which myxobacteria were assigned within at the time). In a similar study into *Bdellovibrio bacteriovorus* (a non-myxobacterial predatory bacterium classified within Deltaproteobacteria at the time), such a pattern of relatedness was taken as evidence a CDS had been acquired by horizontal gene transfer (HGT) from a phylogenetically distant prey organism [20].

The genome sequence of *M. xanthus* DK1622 also revealed the presence of large gene clusters for the synthesis of secondary metabolites (biosynthetic gene clusters—BGCs), which together comprise 8.6% of the genome. This percentage is approximately twice that of *Streptomyces* spp. genomes (which are a similar size to large myxobacterial genomes), suggesting that DK1622 has acquired additional BGCs by HGT [18].

1.2. Other Early Genome Sequences

After that of *M. xanthus* DK1622, the next complete myxobacterial genome to be published was that of *Sorangium cellulosum* So ce56 [21]. The *S. cellulosum* So ce56 genome sequence is 71.4 %GC, typical for myxobacteria, but 13.0 Mbp in length, nearly 4 Mbp larger even than DK1622. One-third of the *S. cellulosum* So ce56 genome is composed of paralogous genes, a lower proportion than for *M. xanthus* DK1622, but with a similar expansion of genes encoding enhancer-binding proteins (EBPs), two-component system (TCS) proteins and Ser/Thr protein kinases [21]. In both *S. cellulosum* So ce56 and *M. xanthus* DK1622, the expansion in number of these protein families is disproportionate to their genome size, with more such proteins per Mbp than any other sequenced bacterial genome at that time. Surprisingly, there was a complete lack of genome-wide synteny (conserved ordering of genes in a genome) observed between the genomes of *S. cellulosum* So ce56 and *M. xanthus* DK1622; however, individual genes exhibited a high degree of local synteny. A total of 1474 of the 9367 CDSs in the *S. cellulosum* So ce56 genome were found in syntenic clusters (mostly corresponding to operons), and the locally syntenic genes also exhibited particularly high sequence conservation with their *M. xanthus* DK1622 counterparts, implying conservation of both genome organisation and function of those genes [21].

Compared to that of *M. xanthus* DK1622, the *S. cellulosum* So ce56 genome has fewer protease and more carbohydrate metabolism genes plus additional genes for nitrogen assimilation [21]. This can be reconciled with *S. cellulosum* So ce56 being a prototroph which can grow on cellulose, while *M. xanthus* DK1622 requires amino acids for growth (as carbon and nitrogen sources), and is auxotrophic for (unable to synthesise) leucine, isoleucine, and valine [18,22].

Also published in 2007 was the draft genome sequence of *Plesiocystis pacifica* SIR-1, sequenced by The Gordon and Betty Moore Foundation Microbial Genome Sequencing project. SIR-1 is described as an aquatic organism with mesophilic salt tolerance and was isolated from a beach on the Japanese Island of Iriomote-jima. Despite being spread over 237 contigs, the genome sequence seems to be close to complete, spanning a total sequence length of 10.6 Mbp. Incomplete genome sequences can be characterised by their N50 and L50 values, where L50 is the minimum number of contigs (x) that together add up to half the total sequence length, and L50 is the length of the xth largest contig. For *P. pacifica* SIR-1, the L50 value is 40 and N50 value is 82,268 bp which are typical values for large myxobacterial genomes. The complete genome of another marine myxobacterium, *Haliangium ochraceum* SMP-2, was published shortly after that of *P. pacifica* SIR-1 [23]. Although myxobacteria have traditionally been viewed as soil bacteria, an increasing number of marine examples have been described. SMP-2 was isolated from seaweed and grows optimally at 2% (w/v) NaCl. As is typical for myxobacteria, it is predatory and forms multicellular fruiting bodies, with a large (9.5 Mbp), high %GC (69.5%) genome [23].

After *H. ochraceum* SMP-2, the next myxobacterial genomes to be completely sequenced belonged to *Corallococcus coralloides* DSM 2259, *Stigmatella aurantiaca* DW4/3-1 and *Myxococcus fulvus* HW-1. All three are typical myxobacterial genome sequences, being large (9–10.3 Mbp), with a high %GC content (>67%), sharing synteny with each other and with *M. xanthus* DK1622 [24–26]. *M. fulvus* HW-1 (reclassified as *Myxococcus macrosporus* HW-1 in October 2018) is a halotolerant marine strain which forms fruiting bodies in low salinity conditions, but which can sporulate without fruiting in saltwater [26]. *C. coralloides* DSM 2259 produces fruiting bodies resembling coral, and it belongs to the most common myxobacterial genus isolated from soils alongside *Myxococcus* [27]. *S. aurantiaca* DW4/3-1 was first sequenced in draft form (released as 579 contigs in Sep 2006), before having its genome completely sequenced four years later [24]. Like *C. coralloides* DSM 2259, *S. aurantiaca* DW4/3-1 also produces morphologically complex fruiting bodies—in this case, with sporangioles mounted on a stalk. Comparisons with the genome sequences of other fruiting myxobacteria showed a lack of conservation of genes involved in fruiting across these myxobacteria, implying the genetic program underlying multicellular development is much more plastic than had been expected [24].

After the publication of the *A. dehalogenans* 2CP-C genome sequence in 2006 [17], genome sequences for a further three members of the genus were made public between 2007 and 2009, with one of those genomes (*Anaeromyxobacter* sp. Fw109-5) subsequently being described in the literature [28]. *Anaeromyxobacter* is an unusual myxobacterial genus as its members do not produce multicellular fruits and they have small genomes for myxobacteria—typically approximately 5 Mbp. *A. dehalogenans* 2CP-C is microaerobic and metabolically versatile, with various gene clusters for electron transport chain complexes acquired by HGT as well as by vertical descent from the ancestral myxobacterium [17]. *Anaeromyxobacter* sp. Fw109-5 is a metal-reducing strain, with the potential for application to the bioremediation of heavy metal-contaminated sites [28].

1.3. Expanding Coverage and Increasing Depth

By 1st July 2012, two draft and ten complete myxobacterial genome sequences were publicly available, from eleven different organisms, spanning eight different genera [19]. Summary statistics of those 12 genome sequences are shown in Table 1. Since then, new sequencing technologies have reduced the cost and increased the accessibility of genome sequencing (see Segerman [29] for a recent review of developments in DNA sequencing technology). As of 2021, commercial services are typically offering 30x coverage of a bacterial genome for less than $70 USD, making genome sequencing an affordable activity even for undergraduate projects and laboratories without large grant incomes. This has resulted in a dramatic exponential increase in the number of available myxobacterial genome sequences (Figure 1).

Table 1. The first 12 published myxobacterial genome sequences (as reported by Huntley et al. [19]), ordered by release date. Sequences without a reference were published variously by: * The Institute for Genomic Research, † The Gordon and Betty Moore Foundation Microbial Genome Sequencing project, or ‡ the United States Department of Energy Joint Genome Institute.

Organism	Mbp	%GC	Contigs	Released	Source	Accession
Anaeromyxobacter dehalogenans 2CP-C	5.0	74.9	1	Jan 2006	[17]	GCA_000013385.1
Myxococcus xanthus DK1622	9.1	68.9	1	Jun 2006	[18]	GCA_000012685.1
Stigmatella aurantiaca DW4/3-1	10.3	67.4	579	Sep 2006	TIGR *	GCA_000168055.1
Plesiocystis pacifica SIR-1^{T}	10.6	70.7	237	Jun 2007	G&BMF MGSP †	GCA_000170895.1
Anaeromyxobacter sp. Fw109-5	5.3	73.5	1	Jul 2007	[28]	GCA_000017505.1
Sorangium cellulosum So ce56	13.0	71.4	1	Nov 2007	[21]	GCA_000067165.1
Anaeromyxobacter sp. K	5.1	74.8	1	Aug 2008	US DOE JGI ‡	GCA_000020805.1
Anaeromyxobacter dehalogenans 2CP-1^{T}	5.0	74.7	1	Jan 2009	US DOE JGI ‡	GCA_000022145.1
Haliangium ochraceum SMP-2^{T}	9.5	69.5	1	Oct 2009	[23]	GCA_000024805.1
Stigmatella aurantiaca DW4/3-1	10.3	67.5	1	Oct 2010	[24]	GCA_000165485.1
Myxococcus macrosporus HW-1	9.0	70.6	1	Jun 2011	[26]	GCA_000219105.1
Corallococcus coralloides DSM 2259^{T}	10.1	69.9	1	Mar 2012	[25]	GCA_000255295.1

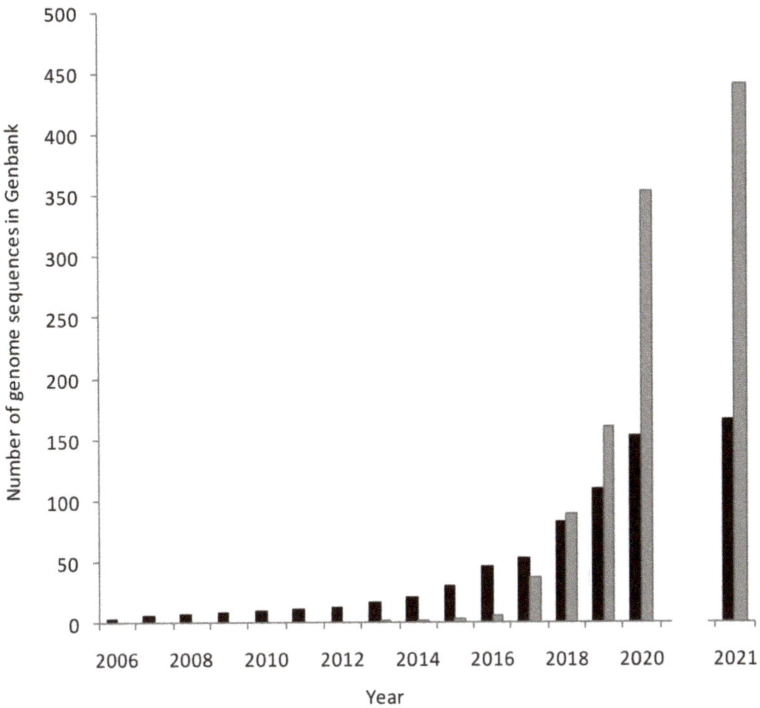

Figure 1. An exponential increase in myxobacterial genome sequencing. The numbers of genome sequences from cultured strains (black columns) and MAGs (grey columns) available at the end of each year are shown. The columns for 2021 only include genomes and MAGs published in the first six months of the year.

It is now possible to routinely sequence newly isolated organisms from the environment, engineered strains, and evolved strains from evolution experiments [30–32]. The increase in genome sequencing activity has provided genome sequences for myxobacterial taxa previously lacking sequenced representatives, and provided more examples of sequenced individuals within key taxa, giving insights into genomic variation within those taxa and the typicality of laboratory model organisms. Figure 2 shows the number of genome sequences currently available for each myxobacterial taxon (as of the 1st July 2021), highlighting a relative dearth of sequences from within families *Vulgatibacteraceae*, *Haliangiaceae*, and *Sandaracinaceae*.

Supplementary Table S1 provides details of all 163 myxobacterial genome sequences deposited in GenBank as of the 1st July 2021, including details of the taxonomy of the organism, the size and %GC of its genome, the number of contigs, date of release and relevant accession numbers. Among the 163 available myxobacterial genomes are 24 draft genomes which we have sequenced and are describing here for the first time (Table 2). Eight of the 24 genomes are from previously undescribed strains, and for those, we also specify where the soil samples were taken which yielded each strain (all in West Wales, UK).

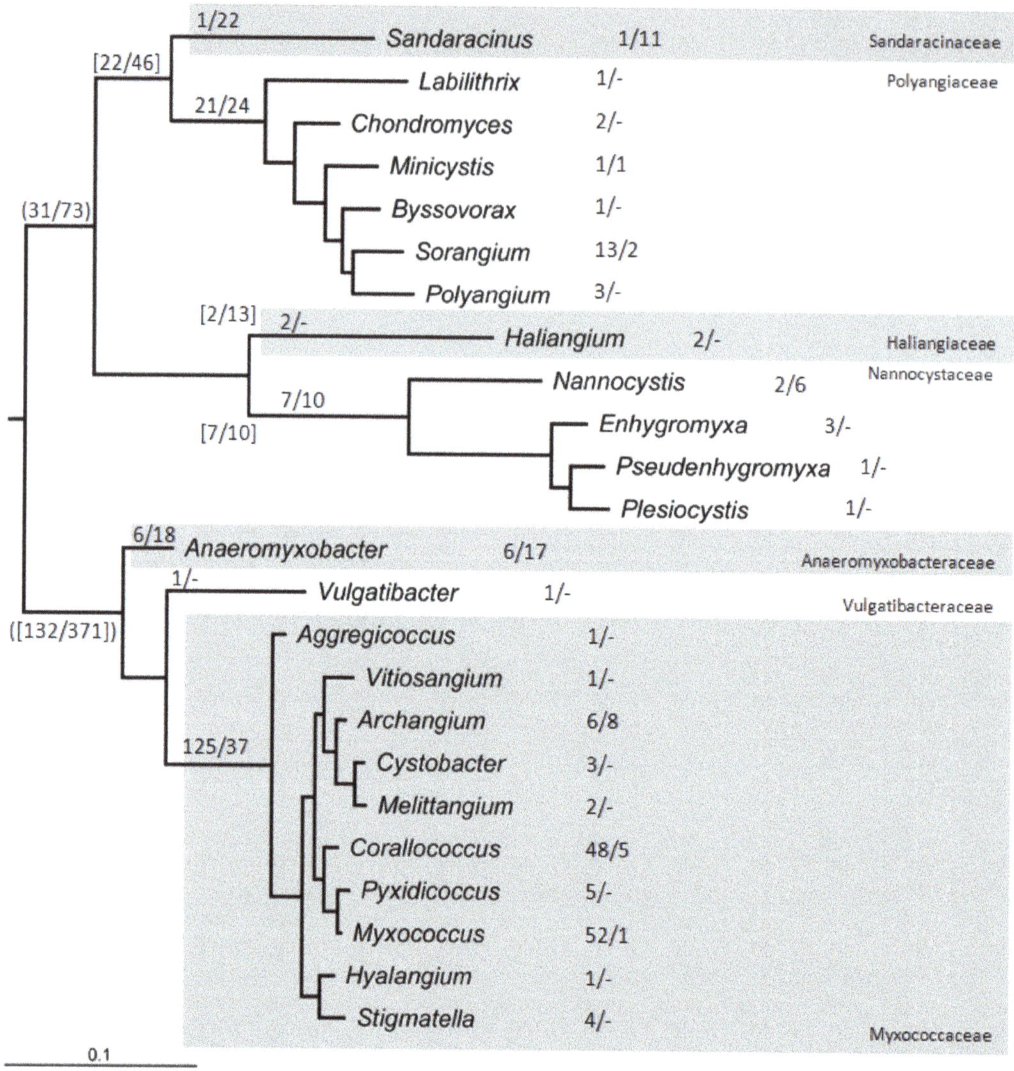

Figure 2. Phylogenetic tree showing the number of genome sequences and MAGs available for sequenced myxobacterial taxa. The tree was produced using 16S rRNA gene sequences from the type strain of each myxobacterial genus (Appendix A). Looking down the tree, families are alternately shaded grey and white. Numbers denote sequenced genomes/MAGs and are shown for each genus, family, order [in square brackets] and class (curved brackets). The Haliangiales and Nanncystales orders each comprise a single family (Haliangiaceae and Nannocystaceae, respectively), while the Myxococcia class contains a single class [Myxococcales]. Not all sequenced organisms/MAGs are taxonomically defined down to the genus, family or order levels (Supplementary Table S1).

Table 2. Summary statistics of 24 genome sequences described here for the first time. The strains are environmental isolates and were assigned to species using genome-based taxonomic principles as described in the text (ANI and dDDH comparisons). Some of the strains (but not their genomes) have been described previously [33].

Strain	Taxonomy	Size (Mbp)	%GC	Contigs	CDS	L50	N50	Coverage	Genbank Accession	Area of Sampling	Reference
CA046B	*Corallococcus carmarthensis*	10.74	69.8	1079	8532	132	23389	71x	JABFJX000000000	Carmarthen	[33]
CA044C	*Corallococcus coralloides*	10.05	70	1379	8124	91	32261	65x	JABFJY000000000	Carmarthen	[33]
AB043B	*Corallococcus exercitus*	10.26	70.2	690	8102	119	26793	77x	JABFJV000000000	Capel Bangor	[33]
CA046A	*Corallococcus exercitus*	9.9	70.5	874	7710	146	21172	77x	JABFJW000000000	Carmarthen	[33]
AB032A	*Corallococcus exiguus*	10.44	69.6	972	8216	132	23729	60x	JABJTS000000000	Penglais Woods, Aberystwyth	[33]
AB038A	*Corallococcus exiguus*	10.57	69.4	631	8281	92	33924	72x	JABJTT000000000	Penglais Woods, Aberystwyth	[33]
AB039A	*Corallococcus exiguus*	10.54	69.4	716	8269	111	28467	91x	JABJTU000000000	Penglais Woods, Aberystwyth	[33]
AM006	*Corallococcus exiguus*	10.59	69.5	865	8306	108	29172	67x	JABNNF000000000	Gogerddan Farm, Aberystwyth	This study
AM007	*Corallococcus exiguus*	10.46	69.6	1264	8357	224	14471	29x	JABNNG000000000	Gogerddan Farm, Aberystwyth	This study
CA046D	*Corallococcus exiguus*	10.5	69.5	955	8310	122	26352	31x	JABNNE000000000	Carmarthen	[33]
AM011	*Myxococcus eversor*	11.62	68.9	688	9104	110	33389	82x	JABXEM000000000	Aberystwyth Harbour	This study
CA033	*Myxococcus llanfairensis* *	11.62	68.8	115	8928	9	312747	17x	JABUMU000000000	Tanerdy Woods, Carmarthen	[33]
CA039A	*Myxococcus llanfairensis* *	11.59	68.7	849	9039	133	27584	59x	JABUMQ000000000	Tanerdy Woods, Carmarthen	[33]
CA040A	*Myxococcus llanfairensis* *	11.72	68.9	70	8992	5	1036580	23x	JABUMR000000000	Tanerdy Woods, Carmarthen	[33]
CA051A	*Myxococcus llanfairensis* *	11.45	68.9	106	8765	7	562936	22x	JABUMS000000000	Llansteffan	[33]
CA056	*Myxococcus llanfairensis* *	11.36	68.9	77	8742	5	658787	26x	JABUMT000000000	Llansteffan	[33]
AM001	*Myxococcus vastator*	9.8	68.8	1946	8912	74	43403	115x	JABXEN000000000	Anglesey	This study
AM009	*Myxococcus vastator*	8.8	70	550	6926	61	45904	138x	JABXEP000000000	Clarach, Ceredigion	This study
AM010	*Myxococcus vastator*	8.93	70	276	6987	34	77358	125x	JABXEO000000000	Gogerddan Farm, Aberystwyth	This study
AB023	*Myxococcus xanthus*	9.13	68.9	221	7181	30	102234	144x	JABFNQ000000000	Gogerddan Farm, Aberystwyth	[33]
AM003	*Myxococcus xanthus*	9.14	69.2	380	7130	57	51156	109x	JABFNS000000000	Anglesey	This study
AM005	*Myxococcus xanthus*	9.15	69.2	413	7148	53	55319	148x	JABFNT000000000	Anglesey	This study
CA029	*Myxococcus xanthus*	9.19	68.8	584	7186	48	56679	93x	JABFNR000000000	Carmarthen	[33]
CA059B	*Pyxidicoccus fallax*	13.39	70.5	1321	10272	232	17466	25x	JABJTR000000000	Llansteffan	[33]

* *Myxococcus llanfairensis* is an abbreviation of *Myxococcus llanfairpwllgwyngyllgogerychwyrndrobwllllantysiliogogogochensis*.

The high coverage of sequencing possible in meta-genome sequencing has also allowed the reconstruction of genomes from uncultured myxobacteria (metagenome-assembled genomes—MAGs). For instance, the draft MAG of *Sandaracinus* sp. NAT8 was reconstructed from metagenomic sequence data generated from ocean samples [34]. The first myxobacterial MAG was added to Genbank in 2013 and within just five years the number of MAGs in Genbank exceeded that of genome sequences from individual strains (Figure 1). Despite their traditional description as soil bacteria, it remains surprising that approximately two-thirds of myxobacterial MAGs originated from aquatic samples—both saline and freshwater (Figure 3), while less than 20% originated from soil samples, although this may reflect bias in selection of sample sources for metagenomic studies rather than having any ecological significance. Supplementary Table S1 also provides details of all 444 MAGs deposited in GenBank as of the 1st July 2021, including the source of the sample which yielded each MAG.

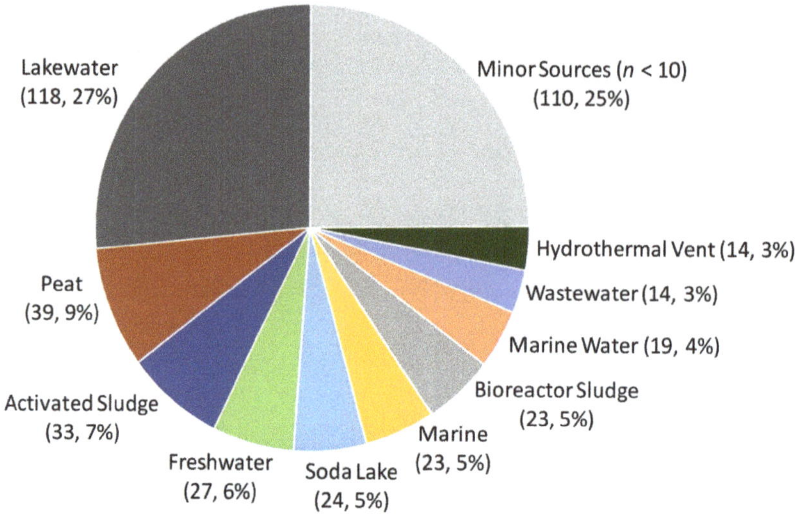

Figure 3. Sources of samples from which myxobacterial 444 MAGs have been derived. The ten sources which have yielded the largest numbers of MAGs are indicated.

Figure 4 shows the relationship between %GC and genome size for myxobacterial genomes and MAGs. Only one of the 163 myxobacterial genome sequences derived from pure strains has a %GC content below 66%, compared to 202 MAGs (46%). Similarly, while 93% of genome sequences from cultured strains have a size above 8.8 Mbp, only 12% of MAGs are that large. It therefore seems highly likely that a large proportion of the 'myxobacterial' MAGs in Genbank are not actually myxobacterial and should be treated with caution.

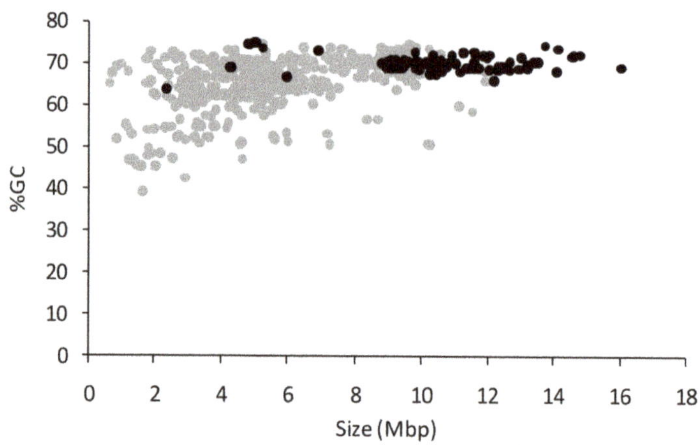

Figure 4. The relationship between genome size (Mbp) and %GC for myxobacterial genome sequences (black) and MAGs (grey).

For the remainder of this paper, when we refer to genome sequences, we only consider those from cultured strains and do not include MAGs unless explicitly stated.

1.4. Genome Sequences and Myxobacterial Classification

In order to understand how genomes evolve as sister lineages diverge, forming new species, genera and families, we need to define the taxonomic relationships between genome-sequenced organisms. Currently, classification of novel myxobacterial taxa requires a polyphasic comparison with pre-existing taxa. Comparators include a variety of phenotypes/properties, typically including fruiting body morphology, colony morphology, cell morphology, nutritional requirements, DNA–DNA hybridisation, optimum growth conditions, fatty acid profiles and enzyme activities [35]. The ability to routinely PCR-amplify and DNA sequence the 16S rRNA gene of organisms led to the inclusion of 16S phylogenetic analysis as a requirement for classification and an objective tool for comparison of large numbers of strains (e.g., [36]). The phylogenetic approach allowed the facile assignment of environmental isolates to individual species. By convention, if the 16S gene sequence of an isolate shares >99% identity with that of the type strain for a species, it can safely be assumed to belong to that species.

Genome sequences are increasingly being used to support taxonomic assignment. DNA–DNA hybridisation (DDH) is an experimental approach, which assesses the sequence similarity of DNA from two sources by measuring the melting temperature of hybridised DNA, and has been used widely in taxonomy. DDH can be calculated directly from genome sequences (as digital DDH or dDDH values) and metrics for genome-wide sequence comparisons have been developed for inter-species and inter-genus comparisons [37]. The ANI (average nucleotide identity) assesses the percentage identity of all genes shared by two genomes, not just the 16S gene, and an ANI value below 95% is good evidence that two genome sequences come from different species [37]. ANI and dDDH-based approaches work equally well on draft and complete genomes. With genome sequences now available for most myxobacterial taxa, it is possible to robustly assign isolates to taxa and identify isolates which might represent novel taxa using their genome sequences alone.

For instance, environmental isolates CA053C, AB025A, AB025B, and AB036A have been described previously and assigned to a genus [33], their genome sequences have also been described [38], but their species membership has not been reported. To define their species, their genome sequences were compared to those of all *Corallococcus* spp. and *Myxococcus* spp. type strains and ANI values calculated. Each strain had an ANI value >95% when compared with only one type strain and no other, allowing assignment to that species (CA053C is *Corallococcus llansteffanensis*, AB025A and AB036A are *M. xanthus*, AB025B is *M. fulvus*). Similarly, Ahearne et al. [39] used ANI to show that *Archangium primigenium* ATCC 29037 is misclassified, actually belonging to a species of *Melittangium*. We have used this approach to identify the species of 37 genome-sequenced strains which were previously only assigned to a genus (indicated in Supplementary Table S1).

There are several advantages to sequence-based taxonomy compared to the polyphasic/16S approach, which can have limited resolution, with no clear criteria for delineating taxa and which can be subject to considerable experimental variation between laboratories [36,40]. In contrast, digital approaches are reproducible, objective, unambiguous and leverage the sequence of every conserved gene. Genome sequence-based approaches have typically supplemented traditional polyphasic approaches in myxobacteria rather than replacing them [41,42]; however, it seems likely that genome sequences will soon be a requirement for definition of a new taxon and may also soon be considered 'type material' [43]. At the moment, type material means cultures of an organism deposited at two international culture collections, thus only culturable organisms can be used to define new taxa. Allowing genome sequences as type material allows taxonomic assignment of uncultured taxa and genomes assembled from metagenomic datasets.

Above the level of the genus, genome sequence analysis has recently allowed a fundamental reclassification of the myxobacteria [44]. Prior to Waite et al.'s study [44], the myxobacteria were classified as order Myxococcales within the class Deltaproteobacteria, and contained three sub-orders (Cystobacterineae, Nannocysterineae and Sorangiineae), containing eight families (Myxococcaceae, Archangiaceae, Vulgatibacteraceae, Phaselicystidaceae, Polyangiaceae, Sandaracinaceae, Nannocystaceae, and Haliangiaceae), seven of which (all but Phaselicystideae) had at least one genome sequence available.

The Waite et al. genome sequence-based reclassification [44] elevated the myxobacteria to form their own phylum (Myxococcota), with two constituent classes (Myxococcia and Polyangia). Class Myxococcia contains a single order (Myxococcales), while class Polyangia contains three (Polyangiales, Nannocystales and Haliangiales). Seven families are distributed between the four orders—the one genus (*Phaselicystis*) within the Phaselicystideae is proposed to be subsumed into Polyangiaceae thereby rendering Phaselicystideae obselete. Archangiaceae is similarly rendered obselete by moving its constituent genera into family Myxococcaceae and a new family (Anaeromyxobacteraceae), which accommodates *Anaeromyxobacter* spp. [44]. Waite et al. [44] also suggest the existence of three further classes (containing four novel orders and five novel families) represented entirely by MAGs. The revised classification provide by Waite et al. [44] is that used in Supplementary Table S1 and Figure 2.

2. Myxobacterial Genome Biology

The availability of genome sequences for defined taxa means it is now possible to investigate how genome properties and gene presence/absence vary within and between taxa. Studies have focussed on gene families with large numbers of members in myxobacterial genomes, genes known to be involved in hallmark myxobacterial behaviours, and how the genome changes during evolution.

2.1. Pan-Genomics

Sequencing the genomes of multiple strains within a single bacterial species revealed that strains often have genes which are lacking in other members of the species. This led to the concept of a species 'pan-genome', which consists of a 'core genome' of genes present in all members of the species, and an 'accessory genome' containing genes which are present in the genomes of some but not all members [45]. 'Core' genes typically include essential/housekeeping genes and genes required for defining properties exhibited by that species, while 'accessory' genes are typically not essential but can confer additional properties on the strains which contain them. Wielgoss et al. [46] found that in *M. xanthus*, gene gain/loss from the accessory genome was faster than amino acid residue substitution rates in core genes by more than an order of magnitude.

Pan-genomes can be considered 'open' or 'closed' [47,48]. In species with closed pan-genomes, individual strains have very similar constituent genomes, entirely composed of core genes and with a very small number of accessory genes [48]. In contrast, members of species with open pan-genomes have relatively small core genomes with a larger proportion of each genome composed of accessory genes. Open pan-genomes continue to increase in size as more genome sequences are considered [47], which can be due to individual members acquiring novel genes by HGT or by lineage specific duplications and diversification.

Figure 5 shows the pan-genome of ten randomly selected strains of *Corallococcus exiguus*, plotting the number of core genes and the total size of the pan-genome as a function of the number of genomes considered. Plots were generated using ROARY [49] and are very similar to those of *M. xanthus* [50]. The core genome falls rapidly from the mean genome size of 8400 genes to 6300 genes as more genomes are considered, showing that on average each strain's genome is composed of 75% core genes and 25% accessory genes. The pan-genome size increases from 8400 genes as more genomes are added, indicating it is an open pan-genome, containing 14,000 genes after 10 genomes are included, i.e., more than half of the pan-genome is composed of accessory genes.

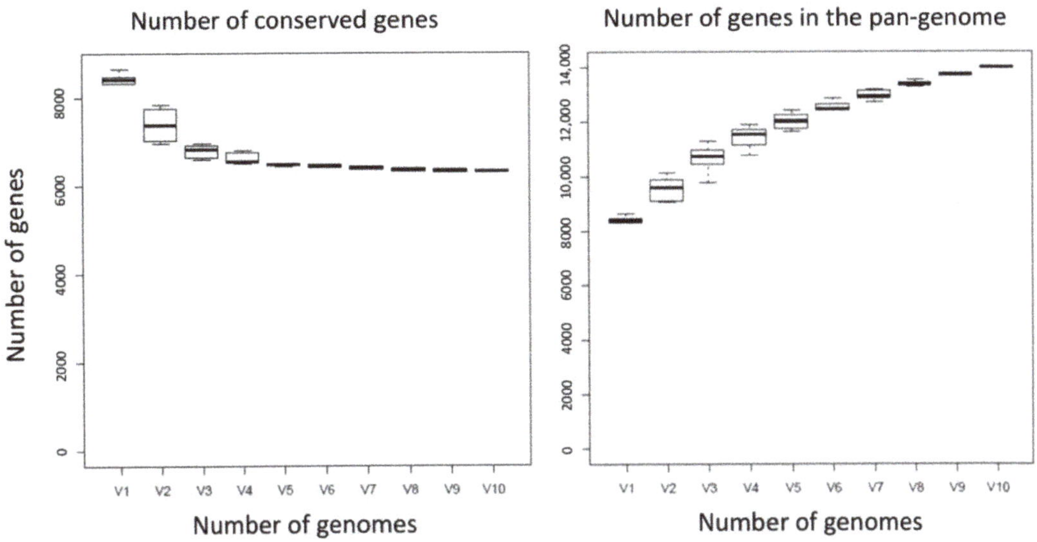

Figure 5. The *Corallococcus exiguus* pan-genome. The number of core genes (**left**) and total number of genes (**right**) for the pan-genome are shown as a function of the number of genomes included (V1–V10). Boxes show the median number of genes ± 1 standard deviation, whiskers show ± 2 standard deviations.

The pan-genome concept can be applied at any taxonomic level—for instance, the core genes of a genus would be those found in all species within that genus, while the accessory genes might be found in just some species within that genus. The pan-genomes of genera *Corallococcus* and *Myxococcus* have also been described [30,42]. At the genus level, the core genomes are much smaller than those of individual species (core genes comprising less than 30% of the average genome), and more than 60% of genes in the accessory genome are found in single species (singletons), indicating considerable variation in gene content between species within each genus [42]. There are also considerable differences between the pan-genomes, despite the constituent species being similarly diverse. The *Corallococcus* pan-genome had three times more core genes than the *Myxococcus* pan-genome which was larger and more open. In both genera BGCs were found to be enriched in the accessory genome, and were found in multiple strains at a much greater frequency than other accessory genome genes, implying they are preferentially conserved and provide a selective advantage [42].

2.2. Comparative Studies—Gene Repertoires

Having genome sequences available has allowed focussed investigations into the genetic repertoire of myxobacteria. Since the first myxobacterial genomes were sequenced, it has been clear that certain types of genes are unusually abundant in myxobacterial genomes, and several publications have now described those gene sets in detail.

BGCs are relatively abundant in myxobacteria and easily identified by sequence homology, with tens of BGCs in most myxobacterial genomes [51]. The metabolites synthesised by BGCs can be predicted by tools such as antiSMASH [52], which reveal a relative abundance of BGCs which produce terpenes, bacteriocins, polyketides and non-ribosomal peptides [51,53]. Several BGCs are associated with cytochrome P450s, which are also abundant in myxobacterial genomes [54]. Most BGCs are thought have been acquired relatively recently by HGT and are part of the accessory genome [18,30,42]. Due to the diversity of their BGCs, it is thought that myxobacteria represent an untapped reservoir of undiscovered natural products, accessible through genome-mining [55–57].

Myxobacterial genomes also contain a very large number of Ser/Thr kinases, which are important components in phosphorylation-dependent signalling pathways, predominantly found in eukaryotes [58]. Despite being generally uncommon in bacteria, individual myxobacterial genomes can encode more than a hundred Ser/Thr kinases, often with unusual combinations of domains for signal transduction (including protein-protein interaction domains, ligand-binding domains and additional kinase/phosphatase domains). The large numbers of kinases are disproportionate with genome size as a consequence of extensive gene duplication [58]. As might be expected, the number of genes encoding Ser/Thr phospho-protein phosphatases in myxobacterial genomes correlates with their number of Ser/Thr kinases, but in far smaller numbers [59].

TCS signalling pathways are the dominant signalling pathways in bacteria and are comprised of a sensor kinase protein which autophosphorylates upon stimulation and then transfers the phosphoryl group to a partner response regulator (RR). Large myxobacterial genomes typically contain more than 250 TCS genes, which is far more than would be expected even for their large genome sizes [60,61]. Myxobacterial TCS genes are often found in complex clusters of TCS genes, or encode TCS proteins with multiple TCS domains, implying a large degree of signal integration between TCS pathways. RRs usually have effector domains in addition to their phosphorylatable 'receiver' domain, but a large proportion of myxobacterial RRs lack such domains, implying they regulate downstream processes via protein-protein interactions [60].

CheY is an example of such a receiver-only RR, which receives phosphoryl groups from the kinase CheA in the TCS signalling pathway governing chemotaxis in diverse organisms. Myxobacterial motility is mechanistically complicated [6], with two distinct engines giving rise to two modes of motility—single-celled 'adventurous' motion (A-motility), or communal 'social' movement (S-motility). Myxobacterial genomes encode multiple CheA-CheY 'chemosensory' systems (*M. xanthus* DK1622 has eight), some of which are involved in regulating motility, while others regulate diverse behaviours, including fruiting body development. Some chemosensory systems are conserved broadly across the myxobacteria, while others seem to have been acquired by relatively recent HGT [62].

Prior to the advent of myxobacterial genome sequencing, numerous studies harnessed the power of bacterial genetics to identify large numbers of genes involved with development and/or motility [63]. Having genome sequences then enabled studies into the conservation and universality of those genes within the myxobacteria. For instance, Huntley et al. [24] showed there was substantial commonality between the developmental programs of fruiting myxobacteria, although substantial plasticity in the program was observed when comparing distantly related myxobacteria [19]. Whitworth and Zwarycz [64] found that genes encoding signalling proteins were enriched in the core genome (with virtually all TCS genes being core), and that within the developmental network, plasticity could be observed even within closely related strains.

2.3. Genome Organisation

In addition to the presence/absence of genes in a genome, their relative location and position-dependent properties are also important considerations. For instance, genes of related function are often grouped together into operons under the control of a shared promoter. During DNA replication, genes tend to maintain their relative order on the genome, a property called synteny. However, recombination events, deletion of genes and the incorporation of new genes from duplications or HGT can change the relative order of genes in a genome [65].

Huntley et al. [19] assessed 'macro'-synteny across myxobacterial genomes by creating dotplots which mapped the positions of homologues for a pair of genomes. Closely related myxobacterial genomes exhibit a pronounced diagonal line due to synteny (e.g., *M. macrosporus* HW-1 compared with *C. coralloides* DSM 2259). However, some genome comparisons (e.g., comparing *M. xanthus* DK1622 with *S. aurantiaca* DW4/3-1) give X-patterns, which are likely due to symmetric interreplichore inversions. Such inversions are the result of recombination between DNA at replication forks, which proceed bidirectionally around the circular chromosome from the *oriC* origin of replication to the *ter* terminus [66]. Comparing more distantly related myxobacteria (e.g., *H. ochraceum* SMP-2 compared with *M. xanthus*), gives dotplots which lack any obvious macro-syntenic relationships [19].

Micro-synteny was observed by Pérez et al. [58] in their investigation into the myxobacterial kinome. Genes encoding Ser/Thr kinases often had conserved local context, with neighbouring genes also being found alongside orthologues in other genomes. Ser/Thr kinase genes have been extensively duplicated in some myxobacterial lineages, and the resulting paralogues are often found close to one another in the genome [58]. Similar patterns of local duplication and micro-synteny are also seen for TCS genes [61].

Another common feature of genome organisation is an asymmetric nucleotide composition of the two strands of DNA. This asymmetry is known as GC skew and it inverts along a DNA strand as the strand changes between being the leading or lagging strand [67]. Changes in asymmetry along a chromosome can therefore be exploited to identify the position of the origin *oriC* gene (usually next to the *dnaA* gene) and the terminus *ter* genes. Notably, the *S. cellulosum* So ce56 genome does not display the usual inversion of GC skew, precluding its use to identify *oriC* [21]. However, using a more complex algorithm, it was subsequently suggested that the *oriC* gene was located next to *dnaN*, nearly 2 Mbp from the *dnaA* gene [68].

Some genes and genome properties are not evenly distributed across myxobacterial genomes. For instance, the *S. cellulosum* So ce56 genome sequence has a region between 8.5 Mbp and 12.5 Mbp, spanning the origin, which is enriched for insertion sequences and contains 90% of predicted genomic islands [21]. In *M. xanthus* DK1622, three clustered interspaced short palindromic repeats (CRISPRs) are found clustered close to the origin, while the four rRNA operons seem to occur in two pairs, with members of each pair approximately the same distance from the origin but opposing each other on the chromosome [69]. A non-random distribution of development genes has also been described, with developmental genes involved in intra-cellular signalling being enriched around the origin compared to inter-cellular signalling genes [70]. With all of these observations, it will be interesting to investigate whether such apparently non-random distributions are due to selective pressures based on the functional roles of the genes, whether genomic location affects gene expression/dosage, or whether genomic location is merely a random consequence of evolutionary mechanisms and heritage.

A particularly variable region in the *M. xanthus* genome was found by Wielgoss et al. [46] when they investigated the whole genome sequences of 22 strains exhibiting colony-merger incompatibilities. Compatibility type dictates whether two expanding colonies are able to merge together during growth and the genome sequences of strains from 11 compatibility types revealed four regions which had a high density of SNPs (single nucleotide polymorphisms). For one of the four regions, spanning 150 kbp, the pattern of SNPs matched compatibility type groupings, as did the presence/absence of genes in the region. The region contained prophages (integrated temperate bacteriophages) and several potential toxin genes, prompting suggestions the region dictates compatibility type [46].

A final aspect of genome organisation considered briefly here is the distribution of gene between replicons. While some bacteria contain two chromosomes, all myxobacteria contain single chromosomes and were thought not to harbour plasmids until recently. Plasmids could be introduced into *M. xanthus* but could not be maintained without integrating into the chromosome by homologous recombination, or by integration into a temperate phage *attB* locus [71,72]. The first autonomously replicating myxobacterial plasmid, pMF1, was discovered in *M. fulvus* strain 124B02 [73]. Maintenance of the plasmid is through a toxin–antitoxin system, constituted by a toxic DNA nuclease and a co-transcribed immunity protein [74]. The plasmid contains 23 predicted CDSs; two encode the toxin–antitoxin system, seven are part of the replication and partitioning systems and the remainder have no obvious function, but homologues can be found occasionally in other myxobacterial genomes. The complete genome sequence of *M. fulvus* 124B02 also revealed the presence of paralogues of the pMF1 toxin–antitoxin gene pair in the strain's chromosome [75].

2.4. Genome Evolution

A large number of studies have used contemporary myxobacterial genome sequences to make inferences regarding the evolutionary processes acting on myxobacteria and their genomes. As would be expected, the types of mutation described in myxobacteria are generally seen across the bacteria. For instance, evidence for recombination between replichores, local gene duplications, gene gain/loss, point mutations, small and large insertions/deletions, and gene fusion/fission can all be observed even when just considering the TCS genes of myxobacteria [76].

There is a metabolic burden associated with the replication of genes, and for bacteria which compete by outgrowing one another, there is consequently a selective pressure to reduce their genome sizes by purifying selection [77]. Genes arriving by HGT will mostly be detrimental to the host cell, and liable to be lost quickly unless they confer a selective advantage [78]. For typical bacteria like *Escherichia coli*, with relatively occasional HGT and small genomes, an adaptationist view of bacterial genomes would suggest that the majority of genes in the genome confer a selective advantage and have a functional role. The failure to observe a phenotype upon deleting a gene could be simply explained if the gene had a niche-specific function which could not manifest under laboratory conditions. For organisms with open pan-genomes, genome size will presumably be a consequence of the relative rates of gene gain and gene loss, and therefore genomes with weak streamlining selection would be expected to be larger.

Myxobacteria are slow-growing organisms and seem to out-compete neighbouring organisms by preying upon them rather than outgrowing them. Huntley et al. [19] discuss the trade-off between translational yield and translational robustness, suggesting that myxobacterial evolutionary success is due to metabolic 'efficiency' rather than speed. Presumably, slow but efficient growth results in a weakened pressure to streamline, which coupled with high rates of HGT results in the large size of myxobacterial genomes. (For this reason it would be interesting to determine whether *Anaeromyxobacter* spp. are predatory, given their small genome sizes.) There is a link between slow growth, large genomes and %GC (a GC base pair is more metabolically expensive to produce than an AT base pair), but any causality is yet to be proven [79]. Nevertheless, it is intriguing that although taxonomically distant from the myxobacteria, the slow-growing, high %GC, BGC-replete members of genus *Streptomyces* have recently been shown to have widespread predatory activity too [80].

A lack of streamlining selection would allow accessory genomes to accumulate genes which do not necessarily confer an immediate fitness advantage, but are instead contingent—for instance, potentially providing a future selective advantage. In the context of multicellular development, reduced streamlining may provide more time for strains to assimilate newly acquired regulators into the developmental program before they are lost from the genome, increasing plasticity and the potential for regulatory innovation [64]. In the context of predation, this may allow maintenance of a diverse arsenal of potentially useful weapons—a sensible strategy considering the inevitability of resistance evolution in prey organisms, and which chimes with the broad prey range exhibited by myxobacterial predators [38].

Nair et al. [81] investigated genome changes in co-evolving co-cultures of *M. xanthus* and *E. coli*. They found reciprocal adaptation between the predator and prey, stimulation of mutation rates and the emergence of mutator genotypes. It would seem that despite taking a generalist approach to predation, myxobacteria can also evolve to increase their predation of particular prey, and that predation per se can drive innovation. Predation could also stimulate innovation through HGT of genes into predator genomes from DNA released by their lysed prey, although genomic signatures of such events are elusive [18].

Nevertheless, HGT from non-myxobacteria would seem to be a major driver for the evolution of myxobacterial accessory genomes: most genes in the accessory genomes of myxobacterial species are singletons (i.e., found only in single genomes), and little exchange is observed between myxobacteria, except between closely related strains [38,46]. Rates of gene gain and loss are high relative to the rate of speciation, yet sequence-based evidence for HGT (e.g., regions with anomalous GC skew or %GC), is missing from myxobacterial genomes [18,19]. Either newly acquired genes are converted to resemble the host genome very quickly (a process called amelioration), or there is selection such that only 'myxobacterial-like' sections of DNA are effectively retained/integrated.

Myxobacteria can take up foreign DNA by transformation and transduction, but conjugation has not been observed. *M. xanthus* is naturally competent and has been shown to acquire drug-resistance genes from other bacteria [82,83]. Relevant to transduction, several temperate bacteriophages of *Myxococcus* spp. have been identified, and various strains of *M. xanthus* carry prophages of *Mx alpha* in their genomes [84]. The prophages reside within the variable region identified by Wielgoss et al. [46] that is responsible for colony merger compatibility and they contain toxin/antitoxin systems responsible for kin discrimination [85]. The incorporation of viral and other incoming DNA into the myxobacterial genome is likely to depend upon the activity of CRISPR-Cas systems, and in *M. xanthus* DK1622 two of the three CRISPR-Cas systems are involved in another social phenomenon—multicellular development [84]. In the original Genbank annotation of the DK1622 genome, 27 CDSs spread over eight loci were annotated as phage proteins, including six recombinases (integrases/excisionases). The *M. xanthus* DK1622 genome also encodes 53 transposases, belonging to seven different IS (insertion sequence) families, suggesting that myxobacterial genomes are shaped by the frequent passage of mobile genetic elements.

2.5. Comparative Studies—Evolution of Specific Myxobacterial Systems

Many studies have investigated the evolution of particular myxobacterial genes and behaviours by comparative analysis of extant genes. The examples below are illustrative rather than comprehensive, but give an idea of the breadth of research activity.

Goldman et al. [86] investigated the evolution of fruiting body formation, finding that three-quarters of developmental genes were inherited vertically. Looking across the *M. xanthus* DK1622 genome, they also identified frequent acquisition of metabolic genes by HGT, including several components of the electron transport chain, reminiscent of the observations of Thomas et al. [17] for *A. dehalogenans* 2CP-C. Other examples of myxobacterial genes gained by HGT include those encoding sterol biosynthesis, an unusual phenomenon in bacteria, which myxobacteria likely acquired from eukaryotes [87].

Other studies have investigated the origin of genes which appear to have arisen de novo within myxobacterial evolution. The Pxr non-coding RNA which regulates fruiting body formation seems to have evolved within the Cystobacterineae sub-order (now order Myxococcales), while the devI regulator of fruiting seems to be a very recent innovation within *M. xanthus* [88,89]. Sequence analysis of 120 strains isolated from six fruiting bodies has shown that genomic changes are concentrated in 'selection hot-spots' and also characterised the rate of endemic diversification [32].

Luciano et al. [90] used a phylogenomic approach to characterise the evolution of candidate genes potentially involved in gliding motility. Using evolutionary and synteny-based arguments they identified three genetic clusters encoding basal motility machinery. Their results also suggested a model for the evolution of gliding motility wherein a core set of ancestral genes of unknown function subsequently recruited extra functional modules [90]. A similar mode of evolution has also been suggested for the type IV pili-based motility systems of myxobacteria [91].

It is also worth noting here an intriguing hypothesis concerning myxobacterial evolution, which suggests that an ancestral myxobacterium may have evolved into a non-myxobacterium. The syntrophy hypothesis proposes that the eukaryotic common ancestor was the result of a tripartite symbiosis involving a myxobacterium-like deltaproteobacterium, which became the eukaryotic cytoplasm [92]. The hypothesis suggests the involvement of a myxobacterial-like organism due to many features of myxobacterial biology which are unusual for bacteria, but common to eukaryotes, including (among many examples) defensins, eukaryotic-like Ser/Thr kinases and enhanceosomes [58,93,94].

3. Myxobacterial Post-Genomics

The availability of a genome sequence is a pre-requisite for several 'omics technologies, particularly transcriptome and proteome analyses. The widespread application of such approaches to myxobacteria has led to the generation of large numbers of 'omics datasets, albeit mainly for *M. xanthus*. Increasingly, 'omics studies and other post-genomic approaches are providing holistic insights into myxobacterial taxonomy, evolution and molecular biology.

3.1. Molecular Genetics

The availability of a genome sequence can inform us about the function and origin of its constituent genes via comparative genomics analyses and it allows the directed study of individual genes or sets of gene in that genome (e.g., [95–97]. The roles of genes can be inferred if they share homology with genes of known function in other organisms, but comparative genomics also makes possible the identification of candidate genes with no obvious functional relationship with the role, including those encoding hypothetical proteins [98]. For example, Luciano et al. [90] made functional predictions of gliding motility genes using synteny-based arguments, while Sutton et al. [38] correlated gene presence/absence with predatory activity to identify candidate predation genes. Genome re-sequencing of spontaneous mutants or members of mutant libraries also allows the identification of genes responsible for the phenotypes under study, a strategy exemplified by the identification of the Pxr non-coding RNA [99].

There is a well-established genetic toolbox for *M. xanthus* that can be used to investigate gene function experimentally, reviewed by Murphy and Garza [9]. Until recently, plasmids that could replicate in myxobacteria were unknown, but plasmids could be introduced into myxobacteria as suicide vectors. Integration into the chromosome could be engineered to happen by homologous recombination with PCR-cloned DNA, or by inclusion of a phage integrase gene and attachment site to direct insertion into the chromosomal *attB* site (e.g., [100,101]. Single recombination of a plasmid containing an internal fragment of a gene could result in insertional disruption of the target gene, while the use of counter selectable makers such as *galK*, allowed the creation of in-frame unmarked deletions via double recombination (e.g., [102]). Since Murphy and Garza's review, the genetic toolbox for myxobacteria has expanded significantly. A replicative plasmid was discovered in *M. fulvus* 124B02, which could be maintained in *M. xanthus* [73]. Inducible promoter systems have been developed, with induction by IPTG (isopropyl-β-D-thiogalactopyranoside), copper or vanillate [103,104]. More recently, genome editing systems have been developed for *M. xanthus*, which would not have been possible without the availability of its genome sequence. A Cre-Lox recombination system was used to engineer a 244 Kbp deletion in DK1622 [105], while CRISPR/Cas-based systems have been used to delete large BGC fragments and to stimulate the expression of a BGC heterologously expressed in *M. xanthus* [106,107].

3.2. Transcriptomics

Prior to the advent of genome sequencing, transcription was normally assayed in *M. xanthus* using northern blots or reporter genes. Typically, a promoterless *lacZ* gene would be cloned downstream of a promoter and introduced into the chromosome (typically at the *attB* site), with production of LacZ (measured by colorimetric assays) indicating transcriptional activity (e.g., [108]). Alternatively, a promoterless *lacZ* within the transposon Tn*5 lac* allowed transcription to be assessed wherever a transposon had inserted into the chromosome [109]. Reverse transcriptase PCR (RT-PCR) increased the ease with which transcriptional assays of sequenced genes could be performed [110], and the arrival of the *M. xanthus* genome sequence allowed the approach to be applied to any gene in the chromosome.

Having the genome sequence of *M. xanthus* DK1622 also allowed the entire transcriptome to be assessed simultaneously, using microarrays initially, and then RNA-seq. The first microarrays were used to study EBPs (enhancer-binding proteins), which are a family of regulatory proteins which stimulate transcription from Sigma54-dependent promoters [95]. Fragments of 371 putative CDSs in the M1 genome sequence were amplified by PCR and spotted onto glass slides. In a two-colour experiment, RNA is extracted from cells grown in (e.g.,) two different conditions, at different time-points during development or from different mutant backgrounds. The RNA samples are reverse-transcribed into cDNA and the two samples of cDNA labelled with different fluorescent dyes. The labelled cDNA samples are then mixed and hybridised to DNA spots on the microarray slide. Up- or down-regulation of CDSs is observed based on the relative intensity of the two fluorescent dyes at each spot. Jakobsen et al. [95], used this approach to investigate changes in transcription during development, identifying 11 transcriptional regulators and six EBPs which were induced 12 h into fruiting body formation.

The first 'genome-wide' microarrays were developed by the *Myxococcus* Microarray Consortium and included spots for 88% of the CDSs in the *M. xanthus* DK1622 genome. Several studies used the arrays to identify which genes are regulated by transcriptional regulators, by comparing gene expression profiles of wild-type strains with those of strains carrying a mutation in the transcriptional regulator gene. For example, Diodati et al. [111], used the arrays to investigate the function of Nla18, a developmental EBP, by comparing transcription in wild-type cells with that of an *nla18* mutant. Surprisingly, in addition to developmental genes, >700 genes were differentially expressed during vegetative growth in the *nla18* mutant compared to the wild-type. Other regulators studied in this way included the response regulators DigR and PhoP4 and the non-coding RNA Pxr [112–114]. Bode et al. [115] used a similar approach to investigate the synthesis of isovaleryl-CoA by assessing transcriptional changes in *bkd* mutants, which are unable to synthesis isovaleryl-CoA via the branched-chain keto acid dehydrogenase complex. Genes identified as being up-regulated in *bkd* mutants included genes encoding an alternative pathway of isovaleryl-CoA synthesis (from 3-hydroxy-3-methylglutaryl-CoA).

Other studies used the microarrays to assess global patterns of gene expression changes associated with a particular biological process in wild-type strains. Shi et al. [116] investigated two-component system genes encoded in the *M. xanthus* DK1622 genome, and assessed which were differentially expressed during fruiting body formation. To help disentangle the regulation of sporulation from that of fruiting body formation, Müller et al. [117] undertook transcriptome profiling of glycerol-induced sporulation, which occurs in single cells without the normal requirement for fruiting body formation. The analysis identified an operon of eight genes up-regulated during sporulation which upon deletion were found to be required for sporulation, but did not affect fruiting body formation. Furusawa et al. [118] investigated differences in expression profiles between yellow and tan phase variants of *M. xanthus*, identifying 41 genes which were specifically up-regulated in yellow or tan variants, including a gene encoding a transcriptional regulator (HTH-Xre) which was subsequently shown to regulate phase switching.

RNA-seq is a method of transcriptome profiling which directly sequences the cDNA generated from a sample of RNA and maps cDNA reads to CDSs in the genome sequence, with the coverage of reads correlating with relative transcript abundance. It has largely superseded microarray profiling because it sequences all RNAs (including those transcribed from unannotated CDSs and non-coding RNAs), it provides quantitative data for each sample rather than a comparison between samples, and avoids problems associated with hybridisation and probe design [119]. Zhu et al. [120] used a transposon-based system to integrate a heterologous BGC (for the production of epothilones) into the genome of *M.*

xanthus, and strains in which the BGC inserted at different genomic sites were found to produce different quantities of epothilones. To investigate those changes in expression, RNA-seq was used to provide a transcription profile of the entire genome in different insertion strains, and it was discovered that insertion of the BGC at different sites caused selective changes in transcriptional activity across the host genome.

Livingstone et al. [121] used RNA-seq to investigate transcriptome changes during *M. xanthus* predation of *E. coli*. Surprisingly, the presence of live prey significantly induced expression of just 12 genes, despite dead prey inducing expression of >1300 genes. This suggested that myxobacteria do not respond to prey presence per se, instead responding to the nutrients released when prey cells are killed. The RNA-seq approach allowed simultaneous investigation of prey gene expression, revealing the induction of >1500 genes in the prey upon exposure to the predator. Subsequent analysis of the RNA-seq data was also able to identify tens of non-coding RNAs in the *M. xanthus* transcriptome, many of which were differentially regulated by nutrient availability [69]. Subsequently, RNA-seq studies have investigated the regulation of fruiting body formation—variously identifying eight or ten distinct sets of expression profiles [122,123], differentiation of peripheral rods—specialised developmental cell types [124], and genes whose expression was induced by UV light—which included BGCs as well as genes of the LexA/SOS response [125].

3.3. Proteomics

In typical proteomic workflows, proteins are digested with trypsin, the resulting peptides are separated (e.g., by high-performance liquid chromatography) and their mass and 'fragmentation fingerprint' accurately determined using mass spectrometry (MS). This allows the sequence of the peptide to be deduced, which is then matched against a theoretical translation of the CDSs in the relevant genome, identifying and quantifying the protein from which the peptide was derived. In myxobacterial research these approaches were initially applied to proteins which had been separated by polyacrylamide gel electrophoresis—either as bands from one-dimensional gels, or spots from two-dimensional gels. For convenience, an entire lane of a one-dimensional gel could be divided into chunks for analysis, giving a semi-quantitative and low-resolution overview of a whole proteome. This approach was used to characterise the proteomes of the *M. xanthus* inner membrane, outer membrane, outer membrane vesicles (OMVs), and extracellular matrix [126–129].

A similar approach involves 'mapping' a proteome, using two-dimensional gel electrophoresis to separate proteins into discrete spots, and then identifying the proteins within each spot. Comparisons between proteomes can then be undertaken by identifying spots with changes in relative intensity, or by labelling two proteomes with different fluorescent dyes and then mixing them before running them on a single two-dimensional gel. Proteins which were relatively more/less abundant in one of the two proteomes would be highlighted in the map because of their coloration. Dahl et al. [130] used such an approach to investigate the spore proteins of *M. xanthus* and identified three previously unknown sporulation proteins (MspA, MspB and MspC). Spores produced by strains with mutations in the *mspA*, *mspB* and *mspC* genes had an altered cortex layer and were more sensitive to environmental stresses. Chao et al. [131] generated a proteome map for *A. dehalogenans* which included 559 proteins, and used the map to investigate the metabolic shift from growth on fumarate to growth on ferric citrate, which was found to affect the relative abundance of 239 proteins. To investigate the role of the transcriptional regulator ROK, Izzat et al. [132] also used two-dimensional fluorescence difference in-gel electrophoresis to compare the proteomes of wild-type and *rok* mutant strains, identifying 130 proteins which were affected by the *rok* mutation.

Gel-free systems are being increasingly used for proteomics, avoiding problems with labelling, quantification and gel loading. Hot on the heels of the *M. xanthus* and *S. cellulosum* genome sequences becoming available, their proteomes were characterised using gel-free approaches, identifying 631 and 952 proteins, respectively [133,134]. Several proteome studies using gel-free and gel-based approaches have focused on OMVs and other proteomes that have reduced complexity compared to cellular proteomes. Berleman et al. [135] and Whitworth et al. [136] investigated the OMVs of *M. xanthus* DZ2 and *M. xanthus* DK1622, respectively, while Zwarycz et al. [50] assessed proteome variation between the OMVs produced by ten independently isolated strains of *M. xanthus*. Whitworth et al. [136] also characterised the soluble secreted proteins and cytoplasmic proteins of *M. xanthus* DK1622 and found that the composition of the soluble supernatant proteome correlated significantly with that of OMVs, implying that lysis of OMVs may in large part dictate the composition of the soluble secreted proteome.

3.4. Metabolomics and Interactomics

While not relying directly on genomic sequence data, metabolomics studies can be enriched by genome sequences. For instance, Bolten et al. [137] cultivated cells of *S. cellulosum* So ce56 on ^{13}C-labelled glucose and identified the metabolites which incorporated the ^{13}C label using GC/MS (gas chromatography coupled with MS). The authors used the *S. cellulosum* So ce56 genome sequence to construct a model of its metabolic network. This allowed a system-wide inference of metabolic fluxes through the pathways of primary metabolism, identifying glycolysis and the pentose phosphate as the major catabolic pathways, with approximately equal fluxes through both. It was also found that the Entner–Douderoff and glyoxylate pathways were inactive in *S. cellulosum* So ce56, and that 90% of the ATP generated by the TCA cycle was consumed during cell maintenance rather than cell growth [137].

An interactomics approach based around high-throughput DNA sequencing is chromatin immunoprecipitation followed by DNA sequencing (ChIP-seq), in which a DNA-binding protein is cross-linked to its bound DNA, and an antibody used to immunoprecipitate the target protein. Bound DNA is released from the precipitate and sequenced, to reveal which parts of the genome are targeted by the DNA-binding protein. Robinson et al. [138] successfully used this approach on *M. xanthus* to identify 1608 putative binding sites for the developmental regulator MrpC, highlighting its involvement in multiple aspects of the developmental program. Sequence similarities between the 1608 putative binding sites allowed identification of a consensus sequence, which was shown to bind a form of MrpC in vitro.

4. Perspectives

In the 15 years following the publication of the first myxobacterial genome sequences, there has been an explosion in the number of myxobacterial genomes which have been sequenced. This has enabled comparative genomic analyses of diverse aspects of myxobacterial biology and has also made possible the application of post-genomic approaches for systems-level analyses of model myxobacteria. The resulting deluge of data has already provided holistic information about the molecular basis of model myxobacterial behaviours, and many more insights are surely yet to be gleaned from those datasets. Genome sequences and post-genomic datasets have generated numerous hypotheses, which can now be tested using molecular genetics approaches.

While model organisms are invaluable tools for investigating molecular genetics, myxobacterial genomes are highly variable and it is not clear to what extent model myxobacteria represent other members of their taxa. Comparative genomics and identification of homologous genes allow the transfer of knowledge between organisms, but we also need to investigate the functional or evolutionary significance of variations between members of the same taxon. For some myxobacterial taxa, we have tens of sequenced genomes, for other taxa we still have none.

As technology advances, what is currently science fiction can quickly become science fact and as costs decrease, advanced technologies become routinely accessible for greater numbers of scientists. In the near future, we would predict current major challenges in myxobacterial research to be overcome. Perhaps:

- Single-cell transcriptomics will be combined with advanced imaging techniques and single-cell tracking to investigate the epigenetic effects of life history on individuals in a population.
- MAGs will direct efforts to define and cultivate novel taxa which are currently unculturable.
- Genome editing and/or recombineering will be used to produce high-throughput combinatorial gene deletions for investigations into gene function.
- Single amplified genomes will provide insights into evolutionary processes within natural populations.
- Proteomics methods will be used holistically to assess post-translational modifications, particularly those associated with epigenetic regulation of metabolism and signalling.
- Artificial intelligence will be used to integrate multi-omic data and physiological data into systems models and to generate hypotheses for testing.

Supplementary Materials: The following are available online at https://www.mdpi.com/article/10.3390/microorganisms9102143/s1, Table S1: Myxobacterial genome sequences and MAGs in Genbank.

Author Contributions: D.E.W. drafted the manuscript. N.S. performed the pan-genome analysis of *C. exiguus*. E.J.R. compiled genome and MAG datasets. All authors have read and agreed to the published version of the manuscript.

Funding: E.J.R. received a Walter Idris Jones Bursary from Aberystwyth University. N.S. was funded by an AberDoc PhD studentship from Aberystwyth University.

Institutional Review Board Statement: Not applicable.

Informed Consent Statement: Not applicable.

Conflicts of Interest: The authors declare no conflict of interest.

Appendix A

The phylogenetic tree in Figure 2 was generated using the 'one-click' option on the phylogeny.fr webserver. A multiple sequence alignment was generated using MUSCLE and curated with Gblocks. The maximum likelihood tree was constructed using PhyML and rendered with TreeDyn, all with default parameters [139].

References

1. Yang, Z.; Higgs, P.I. *Myxobacteria: Genomics, Cellular and Molecular Biology*; Caister Academic Press: Norfolk, UK, 2014.
2. Whitworth, D.E. *Myxobacteria: Multicellularity and Differentiation*; ASM Press: Washington, DC, USA, 2008.
3. Findlay, B.L. The Chemical Ecology of Predatory Soil Bacteria. *ACS Chem. Biol.* **2016**, *11*, 1502–1510. [CrossRef] [PubMed]
4. Furness, E.; Whitworth, D.E.; Zwarycz, A. Predatory Interactions between Myxobacteria and Their Prey. In *The Ecology of Predation at the Microscale*; Jurkevitch, E., Mitchell, R.J., Eds.; Springer: Cham, Switzerland, 2020; pp. 1–36.
5. Petters, S.; Groß, V.; Söllinger, A.; Pichler, M.; Reinhard, A.; Bengtsson, M.M.; Urich, T. The soil microbial food web revisited: Predatory myxobacteria as keystone taxa? *ISME J.* **2021**, *15*, 2665–2675. [CrossRef]
6. Muñoz-Dorado, J.; Torres, F.J.M.; García-Bravo, E.; Moraleda-Munoz, A.; Pérez, J. Myxobacteria: Moving, Killing, Feeding, and Surviving Together. *Front. Microbiol.* **2016**, *7*, 781. [CrossRef]
7. Kaiser, D. Social gliding is correlated with the presence of pili in *Myxococcus xanthus*. *Proc. Natl. Acad. Sci. USA* **1979**, *76*, 5952–5956. [CrossRef] [PubMed]
8. Kroos, L. Highly Signal-Responsive Gene Regulatory Network Governing Myxococcus Development. *Trends Genet.* **2017**, *33*, 3–15. [CrossRef]
9. Murphy, K.A.; Garza, A.G. Genetic Tools for Studying *Myxococcus xanthus* Biology. In *Myxobacteria: Multicellularity and Differentiation*; Whitworth, D.E., Ed.; ASM Press: Washington, DC, USA, 2014; pp. 491–501.
10. Kuspa, A.; Vollrath, D.; Cheng, Y.; Kaiser, D. Physical mapping of the *Myxococcus xanthus* genome by random cloning in yeast artificial chromosomes. *Proc. Natl. Acad. Sci. USA* **1989**, *86*, 8917–8921. [CrossRef]
11. Zusman, D.R.; Krotoski, D.M.; Cumsky, M. Chromosome replication in *Myxococcus xanthus*. *J. Bacteriol.* **1978**, *133*, 122–129. [CrossRef]
12. Johnson, J.L.; Ordal, E.J. Deoxyribonucleic Acid Homology in Bacterial Taxonomy: Effect of Incubation Temperature on Reaction Specificity. *J. Bacteriol.* **1968**, *95*, 893–900. [CrossRef] [PubMed]

13. Kaiser, D.; Manoil, C.; Dworkin, M. Myxobacteria: Cell Interactions, Genetics, and Development. *Annu. Rev. Microbiol.* **1979**, *33*, 595–639. [CrossRef]
14. Inouye, S.; Ike, Y.; Inouye, M. Tandem repeat of the genes for protein S, a development-specific protein of *Myxococcus xanthus*. *J. Biol. Chem.* **1983**, *258*, 38–40. [CrossRef]
15. Romeo, J.M.; Esmon, B.; Zusman, D.R. Nucleotide sequence of the myxobacterial hemagglutinin gene contains four homologous domains. *Proc. Natl. Acad. Sci. USA* **1986**, *83*, 6332–6336. [CrossRef]
16. Oden, S.; Brocchieri, L. Quantitative frame analysis and the annotation of GC-rich (and other) prokaryotic genomes. An application to *Anaeromyxobacter dehalogenans*. *Bioinformatics* **2015**, *31*, 3254–3261. [CrossRef] [PubMed]
17. Thomas, S.H.; Wagner, R.D.; Arakaki, A.K.; Skolnick, J.; Kirby, J.R.; Shimkets, L.J.; Sanford, R.A.; Löffler, F.E. The Mosaic Genome of *Anaeromyxobacter dehalogenans* Strain 2CP-C Suggests an Aerobic Common Ancestor to the Delta-Proteobacteria. *PLoS ONE* **2008**, *3*, e2103. [CrossRef]
18. Goldman, B.S.; Nierman, W.C.; Kaiser, D.; Slater, S.C.; Durkin, A.S.; Eisen, J.A.; Ronning, C.M.; Barbazuk, W.; Blanchard, M.; Field, C.; et al. Evolution of sensory complexity recorded in a myxobacterial genome. *Proc. Natl. Acad. Sci. USA* **2006**, *103*, 15200–15205. [CrossRef]
19. Huntley, S.; Wuichet, K.; Søgaard-Andersen, L. Genome evolution and content in the myxobacteria. In *Myxobacteria: Genomics, Cellular and Molecular Biology*; Yang, Z., Higgs, P.I., Eds.; Caister Academic Press: Norfolk, UK, 2014; pp. 31–50.
20. Gophna, U.; Charlebois, R.L.; Doolittle, W.F. Ancient lateral gene transfer in the evolution of *Bdellovibrio bacteriovorus*. *Trends Microbiol.* **2006**, *14*, 64–69. [CrossRef] [PubMed]
21. Schneiker, S.; Perlova, O.; Kaiser, O.; Gerth, K.; Alici, A.; O Altmeyer, M.; Bartels, D.; Bekel, T.; Beyer, S.; Bode, E.; et al. Complete genome sequence of the myxobacterium *Sorangium cellulosum*. *Nat. Biotechnol.* **2007**, *25*, 1281–1289. [CrossRef] [PubMed]
22. Bretscher, A.P.; Kaiser, D. Nutrition of *Myxococcus xanthus*, a fruiting myxobacterium. *J. Bacteriol.* **1978**, *133*, 763–768. [CrossRef]
23. Ivanova, N.; Daum, C.; Lang, E.; Abt, B.; Kopitz, M.; Saunders, E.; Lapidus, A.; Lucas, S.; Del Rio, T.G.; Nolan, M.; et al. Complete genome sequence of *Haliangium ochraceum* type strain (SMP-2T). *Stand. Genom. Sci.* **2010**, *2*, 96–106. [CrossRef]
24. Huntley, S.; Hamann, N.; Wegener-Feldbrugge, S.; Treuner-Lange, A.; Kube, M.; Reinhardt, R.; Klages, S.; Müller, R.; Ronning, C.M.; Nierman, W.C.; et al. Comparative Genomic Analysis of Fruiting Body Formation in Myxococcales. *Mol. Biol. Evol.* **2010**, *28*, 1083–1097. [CrossRef]
25. Huntley, S.; Zhang, Y.; Treuner-Lange, A.; Kneip, S.; Sensen, C.W.; Søgaard-Andersen, L. Complete Genome Sequence of the Fruiting Myxobacterium *Corallococcus coralloides* DSM 2259. *J. Bacteriol.* **2012**, *194*, 3012–3013. [CrossRef]
26. Li, Z.-F.; Li, X.; Liu, H.; Liu, X.; Han, K.; Wu, Z.-H.; Hu, W.; Li, F.-F.; Li, Y.-Z. Genome Sequence of the Halotolerant Marine Bacterium *Myxococcus fulvus* HW-1. *J. Bacteriol.* **2011**, *193*, 5015–5016. [CrossRef] [PubMed]
27. Mohr, K.I.; Stechling, M.; Wink, J.; Wilharm, E.; Stadler, M. Comparison of myxobacterial diversity and evaluation of isolation success in two niches: Kiritimati Island and German compost. *MicrobiologyOpen* **2016**, *5*, 268–278. [CrossRef] [PubMed]
28. Hwang, C.; Copeland, A.; Lucas, S.; Lapidus, A.; Barry, K.; del Rio, T.G.; Dalin, E.; Tice, H.; Pitluck, S.; Sims, D.; et al. Complete Genome Sequence of *Anaeromyxobacter* sp. Fw109-5, an Anaerobic, Metal-Reducing Bacterium Isolated from a Contaminated Subsurface Environment. *Genome Announc.* **2015**, *3*, e01449-14. [CrossRef] [PubMed]
29. Segerman, B. The Most Frequently Used Sequencing Technologies and Assembly Methods in Different Time Segments of the Bacterial Surveillance and RefSeq Genome Databases. *Front. Cell. Infect. Microbiol.* **2020**, *10*, 527102. [CrossRef] [PubMed]
30. Livingstone, P.; Morphew, R.; Whitworth, D.E. Genome Sequencing and Pan-Genome Analysis of 23 *Corallococcus* spp. Strains Reveal Unexpected Diversity, with Particular Plasticity of Predatory Gene Sets. *Front. Microbiol.* **2018**, *9*, 3187. [CrossRef]
31. Pogorevc, D.; Panter, F.; Schillinger, C.; Jansen, R.; Wenzel, S.C.; Müller, R. Production optimization and biosynthesis revision of corallopyronin A, a potent anti-filarial antibiotic. *Metab. Eng.* **2019**, *55*, 201–211. [CrossRef]
32. Wielgoss, S.; Wolfensberger, R.; Sun, L.; Fiegna, F.; Velicer, G.J. Social genes are selection hotspots in kin groups of a soil microbe. *Science* **2019**, *363*, 1342–1345. [CrossRef]
33. Livingstone, P.; Morphew, R.; Whitworth, D.E. Myxobacteria Are Able to Prey Broadly upon Clinically-Relevant Pathogens, Exhibiting a Prey Range Which Cannot Be Explained by Phylogeny. *Front. Microbiol.* **2017**, *8*, 1593. [CrossRef]
34. Tully, B.J.; Graham, E.D.; Heidelberg, J.F. The reconstruction of 2,631 draft metagenome-assembled genomes from the global oceans. *Sci. Data* **2018**, *5*, 170203. [CrossRef]
35. Mohr, K.I.; Wolf, C.; Nübel, U.; Szafrańska, A.K.; Steglich, M.; Hennessen, F.; Gemperlein, K.; Kampfer, P.; Martin, K.; Müller, R.; et al. A polyphasic approach leads to seven new species of the cellulose-decomposing genus *Sorangium*, *Sorangium ambruticinum* sp. nov., *Sorangium arenae* sp. nov., *Sorangium bulgaricum* sp. nov., *Sorangium dawidii* sp. nov., *Sorangium kenyense* sp. nov., *Sorangium orientale* sp. nov. and *Sorangium reichenbachii* sp. nov. *Int. J. Syst. Evol. Microbiol.* **2018**, *68*, 3576–3586. [CrossRef]
36. Garcia, R.; Gerth, K.; Stadler, M.; Dogma, I.J.; Müller, R. Expanded phylogeny of myxobacteria and evidence for cultivation of the 'unculturables'. *Mol. Phylogenet. Evol.* **2010**, *57*, 878–887. [CrossRef] [PubMed]
37. Chun, J.; Oren, A.; Ventosa, A.; Christensen, H.; Arahal, D.R.; Da Costa, M.S.; Rooney, A.P.; Yi, H.; Xu, X.-W.; De Meyer, S.; et al. Proposed minimal standards for the use of genome data for the taxonomy of prokaryotes. *Int. J. Syst. Evol. Microbiol.* **2018**, *68*, 461–466. [CrossRef]
38. Sutton, D.; Livingstone, P.; Furness, E.; Swain, M.T.; Whitworth, D.E. Genome-Wide Identification of Myxobacterial Predation Genes and Demonstration of Formaldehyde Secretion as a Potentially Predation-Resistant Trait of *Pseudomonas aeruginosa*. *Front. Microbiol.* **2019**, *10*, 2650. [CrossRef] [PubMed]

39. Ahearne, A.; Albataineh, H.; Dowd, S.; Stevens, D. Assessment of Evolutionary Relationships for Prioritization of Myxobacteria for Natural Product Discovery. *Microorganisms* **2021**, *9*, 1376. [CrossRef]
40. Richter, M.; Rosselló-Móra, R. Shifting the genomic gold standard for the prokaryotic species definition. *Proc. Natl. Acad. Sci. USA* **2009**, *106*, 19126–19131. [CrossRef]
41. Livingstone, P.G.; Ingleby, O.; Girdwood, S.; Cookson, A.R.; Morphew, R.; Whitworth, D.E. Predatory Organisms with Untapped Biosynthetic Potential: Descriptions of Novel *Corallococcus* Species *C. aberystwythensis* sp. nov., *C. carmarthensis* sp. nov., *C. exercitus* sp. nov., *C. interemptor* sp. nov., *C. llansteffanensis* sp. nov., *C. praedator* sp. nov., *C. sicarius* sp. nov. and *C. terminator* sp. nov. *Appl. Environ. Microbiol.* **2020**, *86*, e01931-19. [CrossRef]
42. Chambers, J.; Sparks, N.; Sydney, N.; Livingstone, P.G.; Cookson, A.R.; E Whitworth, D. Comparative Genomics and Pan-Genomics of the Myxococcaceae, including a Description of Five Novel Species: *Myxococcus eversor* sp. nov., *Myxococcus llanfairpwllgwyngyllgogerychwyrndrobwllllantysiliogogogochensis* sp. nov., *Myxococcus vastator* sp. nov., *Pyxidicoccus caerfyrddinensis* sp. nov. and *Pyxidicoccus trucidator* sp. nov. *Genome Biol. Evol.* **2020**, *12*, 2289–2302. [CrossRef]
43. Murray, A.E.; Freudenstein, J.; Gribaldo, S.; Hatzenpichler, R.; Hugenholtz, P.; Kampfer, P.; Konstantinidis, K.T.; Lane, C.E.; Papke, R.; Parks, D.H.; et al. Roadmap for naming uncultivated Archaea and Bacteria. *Nat. Microbiol.* **2020**, *5*, 987–994. [CrossRef] [PubMed]
44. Waite, D.W.; Chuvochina, M.; Pelikan, C.; Parks, D.H.; Yilmaz, P.; Wagner, M.; Loy, A.; Naganuma, T.; Nakai, R.; Whitman, W.B.; et al. Proposal to reclassify the proteobacterial classes Deltaproteobacteria and Oligoflexia, and the phylum Thermodesulfobacteria into four phyla reflecting major functional capabilities. *Int. J. Syst. Evol. Microbiol.* **2020**, *70*, 5972–6016. [CrossRef]
45. Medini, D.; Donati, C.; Tettelin, H.; Masignani, V.; Rappuoli, R. The microbial pan-genome. *Curr. Opin. Genet. Dev.* **2005**, *15*, 589–594. [CrossRef]
46. Wielgoss, S.; Didelot, X.; Chaudhuri, R.; Liu, X.; Weedall, G.; Velicer, G.J.; Vos, M. A barrier to homologous recombination between sympatric strains of the cooperative soil bacterium *Myxococcus xanthus*. *ISME J.* **2016**, *10*, 2468–2477. [CrossRef]
47. Donati, C.; Hiller, N.L.; Tettelin, H.; Muzzi, A.; Croucher, N.; Angiuoli, S.V.; Oggioni, M.; Hotopp, J.C.D.; Hu, F.Z.; Riley, D.R.; et al. Structure and dynamics of the pan-genome of Streptococcus pneumoniae and closely related species. *Genome Biol.* **2010**, *11*, R107. [CrossRef]
48. Argemi, X.; Matelska, D.; Ginalski, K.; Riegel, P.; Hansmann, Y.; Bloom, J.; Pestel-Caron, M.; Dahyot, S.; Lebeurre, J.; Prévost, G. Comparative genomic analysis of Staphylococcus lugdunensis shows a closed pan-genome and multiple barriers to horizontal gene transfer. *BMC Genom.* **2018**, *19*, 621. [CrossRef]
49. Page, A.J.; Cummins, C.A.; Hunt, M.; Wong, V.K.; Reuter, S.; Holden, M.T.G.; Fookes, M.; Falush, D.; Keane, J.A.; Parkhill, J. Roary: Rapid large-scale prokaryote pan genome analysis. *Bioinformatics* **2015**, *31*, 3691–3693. [CrossRef] [PubMed]
50. Zwarycz, A.S.; Livingstone, P.; Whitworth, D.E. Within-species variation in OMV cargo proteins: The *Myxococcus xanthus* OMV pan-proteome. *Mol. Omics* **2020**, *16*, 387–397. [CrossRef] [PubMed]
51. Gemperlein, K.; Zaburannyi, N.; Garcia, R.; La Clair, J.J.; Müller, R. Metabolic and Biosynthetic Diversity in Marine Myxobacteria. *Mar. Drugs* **2018**, *16*, 314. [CrossRef]
52. Blin, K.; Shaw, S.; Kloosterman, A.M.; Charlop-Powers, Z.; van Wezel, G.P.; Medema, M.H.; Weber, T. antiSMASH 6.0: Improving cluster detection and comparison capabilities. *Nucleic Acids Res.* **2021**, *49*, W29–W35. [CrossRef]
53. Moghaddam, J.A.; Crüsemann, M.; Alanjary, M.; Harms, H.; Dávila-Céspedes, A.; Blom, J.; Poehlein, A.; Ziemert, N.; König, G.M.; Schäberle, T.F. Analysis of the Genome and Metabolome of Marine Myxobacteria Reveals High Potential for Biosynthesis of Novel Specialized Metabolites. *Sci. Rep.* **2018**, *8*, 16600. [CrossRef] [PubMed]
54. Khatri, Y.; Hannemann, F.; Perlova, O.; Müller, R.; Bernhardt, R. Investigation of cytochromes P450 in myxobacteria: Excavation of cytochromes P450 from the genome of *Sorangium cellulosum* So ce56. *FEBS Lett.* **2011**, *585*, 1506–1513. [CrossRef]
55. Burgard, C.; Zaburannyi, N.; Nadmid, S.; Maier, J.; Jenke-Kodama, H.; Luxenburger, E.; Bernauer, H.S.; Wenzel, S.C. Genomics-Guided Exploitation of Lipopeptide Diversity in Myxobacteria. *ACS Chem. Biol.* **2017**, *12*, 779–786. [CrossRef]
56. Gregory, K.; Salvador, L.A.; Akbar, S.; Adaikpoh, B.; Stevens, D.C. Survey of Biosynthetic Gene Clusters from Sequenced Myxobacteria Reveals Unexplored Biosynthetic Potential. *Microorganisms* **2019**, *7*, 181. [CrossRef]
57. Hug, J.J.; Panter, F.; Krug, D.; Müller, R. Genome mining reveals uncommon alkylpyrones as type III PKS products from myxobacteria. *J. Ind. Microbiol. Biotechnol.* **2019**, *46*, 319–334. [CrossRef]
58. Perez, J.; Castaneda-Garcia, A.; Jenke-Kodama, H.; Muller, R.; Munoz-Dorado, J. Eukaryotic-like protein kinases in the prokaryotes and the myxobacterial kinome. *Proc. Natl. Acad. Sci. USA* **2008**, *105*, 15950–15955. [CrossRef]
59. Treuner-Lange, A. The Phosphatomes of the Multicellular Myxobacteria *Myxococcus xanthus* and *Sorangium cellulosum* in Comparison with Other Prokaryotic Genomes. *PLoS ONE* **2010**, *5*, e11164. [CrossRef] [PubMed]
60. Whitworth, D.E.; Cock, P.J.A. Two-component systems of the myxobacteria: Structure, diversity and evolutionary relationships. *Microbiology* **2008**, *154*, 360–372. [CrossRef] [PubMed]
61. Whitworth, D.E.; Cock, P.J.A. Myxobacterial Two-Component Systems. In *Myxobacteria: Multicellularity and Differentiation*; Whitworth, D.E., Ed.; ASM Press: Washington, DC, USA, 2008; pp. 169–189.
62. Sharma, G.; Khatri, I.; Subramanian, S. Comparative Genomics of Myxobacterial Chemosensory Systems. *J. Bacteriol.* **2018**, *200*, e00620-17. [CrossRef]
63. Kaiser, D. Signaling in myxobacteria. *Annu. Rev. Microbiol.* **2004**, *58*, 75–98. [CrossRef]

64. Whitworth, D.E.; Zwarycz, A. A Genomic Survey of Signalling in the Myxococcaceae. *Microorganisms* **2020**, *8*, 1739. [CrossRef] [PubMed]
65. Bentley, S.D.; Parkhill, J. Comparative Genomic Structure of Prokaryotes. *Annu. Rev. Genet.* **2004**, *38*, 771–791. [CrossRef]
66. Khedkar, S.; Seshasayee, A.S.N. Comparative Genomics of Interreplichore Translocations in Bacteria: A Measure of Chromosome Topology? *G3* **2016**, *6*, 1597–1606. [CrossRef]
67. Mackiewicz, P. Where does bacterial replication start? Rules for predicting the oriC region. *Nucleic Acids Res.* **2004**, *32*, 3781–3791. [CrossRef] [PubMed]
68. Gao, F.; Zhang, C.-T. Origins of Replication in *Sorangium cellulosum* and *Microcystis aeruginosa*. *DNA Res.* **2008**, *15*, 169–171. [CrossRef] [PubMed]
69. Whitworth, D.E.; Swain, M.T. A survey of non-coding RNAs in the social and predatory myxobacterium *Myxococcus xanthus* DK1622. *Mol. Omics* **2020**, *16*, 492–502. [CrossRef] [PubMed]
70. Evans, A.R.; Whitworth, D.E. Correlations between the role, sequence conservation, genomic location and severity of phenotype in myxobacterial developmental genes. *FEMS Microbiol. Lett.* **2010**, *312*, 40–45. [CrossRef] [PubMed]
71. Saulnier, P.; Hanquier, J.; Jaoua, S.; Reichenbach, H.; Guespin-Michel, J.F. Utilization of IncP-1 Plasmids as Vectors for Transposon Mutagenesis in Myxobacteria. *Microbiology* **1988**, *134*, 2889–2895. [CrossRef]
72. Magrini, V.; Creighton, C.; Youderian, P. Site-Specific Recombination of Temperate *Myxococcus xanthus* Phage Mx8: Genetic Elements Required for Integration. *J. Bacteriol.* **1999**, *181*, 4050–4061. [CrossRef]
73. Zhao, J.-Y.; Zhong, L.; Shen, M.-J.; Xia, Z.; Cheng, Q.-X.; Sun, X.; Zhao, G.-P.; Li, Y.; Qin, Z.-J. Discovery of the Autonomously Replicating Plasmid pMF1 from *Myxococcus fulvus* and Development of a Gene Cloning System in *Myxococcus xanthus*. *Appl. Environ. Microbiol.* **2008**, *74*, 1980–1987. [CrossRef]
74. Li, Y.-J.; Liu, Y.; Zhang, Z.; Chen, X.-J.; Gong, Y.; Li, Y. A Post-segregational Killing Mechanism for Maintaining Plasmid PMF1 in Its *Myxococcus fulvus* Host. *Front. Cell. Infect. Microbiol.* **2018**, *8*, 274. [CrossRef]
75. Chen, X.-J.; Han, K.; Feng, J.; Zhuo, L.; Li, Y.-J.; Li, Y.-Z. The complete genome sequence and analysis of a plasmid-bearing myxobacterial strain *Myxococcus fulvus* 124B02 (M 206081). *Stand. Genom. Sci.* **2016**, *11*, 1. [CrossRef]
76. Whitworth, D.E. Genome-wide analysis of myxobacterial two-component systems: Genome relatedness and evolutionary changes. *BMC Genom.* **2015**, *16*, 780. [CrossRef]
77. Giovannoni, S.J.; Thrash, C.; Temperton, B. Implications of streamlining theory for microbial ecology. *ISME J.* **2014**, *8*, 1553–1565. [CrossRef] [PubMed]
78. Baltrus, D.A. Exploring the costs of horizontal gene transfer. *Trends Ecol. Evol.* **2013**, *28*, 489–495. [CrossRef]
79. Almpanis, A.; Swain, M.; Gatherer, D.; McEwan, N. Correlation between bacterial G+C content, genome size and the G+C content of associated plasmids and bacteriophages. *Microb. Genom.* **2018**, *4*, e000168. [CrossRef]
80. Kumbhar, C.; Mudliar, P.; Bhatia, L.; Kshirsagar, A.; Watve, M. Widespread predatory abilities in the genus Streptomyces. *Arch. Microbiol.* **2014**, *196*, 235–248. [CrossRef]
81. Nair, R.R.; Vasse, M.; Wielgoss, S.; Sun, L.; Yu, Y.-T.N.; Velicer, G.J. Bacterial predator-prey coevolution accelerates genome evolution and selects on virulence-associated prey defences. *Nat. Commun.* **2019**, *10*, 1–10. [CrossRef]
82. Parish, J.H. Transfer of Drug Resistance to Myxococcus from Bacteria Carrying Drug-resistance Factors. *J. Gen. Microbiol.* **1975**, *87*, 198–210. [CrossRef] [PubMed]
83. Wang, J.; Hu, W.; Lux, R.; He, X.; Li, Y.; Shi, W. Natural Transformation of *Myxococcus xanthus*. *J. Bacteriol.* **2011**, *193*, 2122–2132. [CrossRef] [PubMed]
84. Vasse, M.; Wielgoss, S. Bacteriophages of *Myxococcus xanthus*, a Social Bacterium. *Viruses* **2018**, *10*, 374. [CrossRef]
85. Vassallo, C.N.; Cao, P.; Conklin, A.; Finkelstein, H.; Hayes, C.S.; Wall, D. Infectious polymorphic toxins delivered by outer membrane exchange discriminate kin in myxobacteria. *eLife* **2017**, *6*, 29397. [CrossRef]
86. Goldman, B.; Bhat, S.; Shimkets, L.J. Genome Evolution and the Emergence of Fruiting Body Development in *Myxococcus xanthus*. *PLoS ONE* **2007**, *2*, e1329. [CrossRef]
87. Desmond, E.; Gribaldo, S. Phylogenomics of Sterol Synthesis: Insights into the Origin, Evolution, and Diversity of a Key Eukaryotic Feature. *Genome Biol. Evol.* **2009**, *1*, 364–381. [CrossRef]
88. Chen, I.-C.K.; Griesenauer, B.; Yu, Y.-T.N.; Velicer, G.J. A recent evolutionary origin of a bacterial small RNA that controls multicellular fruiting body development. *Mol. Phylogenet. Evol.* **2014**, *73*, 1–9. [CrossRef] [PubMed]
89. Rajagopalan, R.; Wielgoss, S.; Lippert, G.; Velicer, G.J.; Kroos, L. devIls an Evolutionarily Young Negative Regulator of *Myxococcus xanthus* Development. *J. Bacteriol.* **2015**, *197*, 1249–1262. [CrossRef]
90. Luciano, J.; Agrebi, R.; Le Gall, A.V.; Wartel, M.; Fiegna, F.; Ducret, A.; Brochier-Armanet, C.; Mignot, T. Emergence and Modular Evolution of a Novel Motility Machinery in Bacteria. *PLoS Genet.* **2011**, *7*, e1002268. [CrossRef]
91. Sharma, G.; Burrows, L.L.; Singer, M. Diversity and Evolution of Myxobacterial Type IV Pilus Systems. *Front. Microbiol.* **2018**, *9*, 1630. [CrossRef]
92. López-García, P.; Moreira, D. The Syntrophy hypothesis for the origin of eukaryotes revisited. *Nat. Microbiol.* **2020**, *5*, 655–667. [CrossRef] [PubMed]
93. Zhu, S. Evidence for myxobacterial origin of eukaryotic defensins. *Immunogenetics* **2007**, *59*, 949–954. [CrossRef]
94. Elías-Arnanz, M.; Padmanabhan, S.; Murillo, F.J. The regulatory action of the myxobacterial CarD/CarG complex: A bacterial enhanceosome? *FEMS Microbiol. Rev.* **2010**, *34*, 764–778. [CrossRef] [PubMed]

95. Jakobsen, J.S.; Jelsbak, L.; Jelsbak, L.; Welch, R.D.; Cummings, C.; Goldman, B.; Stark, E.; Slater, S.; Kaiser, D. σ 54 Enhancer Binding Proteins and *Myxococcus xanthus* Fruiting Body Development. *J. Bacteriol.* **2004**, *186*, 4361–4368. [CrossRef] [PubMed]
96. Whitworth, D.E.; Millard, A.; Hodgson, D.A.; Hawkins, P.F. Protein-protein interactions between two-component system transmitter and receiver domains of *Myxococcus xanthus*. *Proteomics* **2008**, *8*, 1839–1842. [CrossRef]
97. Vassallo, C.N.; Wall, D. Self-identity barcodes encoded by six expansive polymorphic toxin families discriminate kin in myxobacteria. *Proc. Natl. Acad. Sci. USA* **2019**, *116*, 24808–24818. [CrossRef]
98. Whitworth, D.E. Genomes and knowledge—A questionable relationship? *Trends Microbiol.* **2008**, *16*, 512–519. [CrossRef]
99. Velicer, G.J.; Raddatz, G.; Keller, H.; Deiss, S.; Lanz, C.; Dinkelacker, I.; Schuster, S.C. Comprehensive mutation identification in an evolved bacterial cooperator and its cheating ancestor. *Proc. Natl. Acad. Sci. USA* **2006**, *103*, 8107–8112. [CrossRef] [PubMed]
100. Caberoy, N.B.; Welch, R.D.; Jakobsen, J.S.; Slater, S.C.; Garza, A.G. Global Mutational Analysis of NtrC-Like Activators in *Myxococcus xanthus*: Identifying Activator Mutants Defective for Motility and Fruiting Body Development. *J. Bacteriol.* **2003**, *185*, 6083–6094. [CrossRef] [PubMed]
101. Whitworth, D.E.; Bryan, S.J.; Berry, A.; McGowan, S.; Hodgson, D.A. Genetic Dissection of the Light-Inducible carQRS Promoter Region of *Myxococcus xanthus*. *J. Bacteriol.* **2004**, *186*, 7836–7846. [CrossRef] [PubMed]
102. Ueki, T.; Inouye, S.; Inouye, M. Positive-negative KG cassettes for construction of multi-gene deletions using a single drug marker. *Gene* **1996**, *183*, 153–157. [CrossRef]
103. Iniesta, A.A.; García-Heras, F.; Abellón-Ruiz, J.; García, A.G.; Elías-Arnanz, M. Two Systems for Conditional Gene Expression in *Myxococcus xanthus* Inducible by Isopropyl-β- d -Thiogalactopyranoside or Vanillate. *J. Bacteriol.* **2012**, *194*, 5875–5885. [CrossRef]
104. Santos, N.G.; Treuner-Lange, A.; Moraleda-Munoz, A.; García-Bravo, E.; García-Hernández, R.; Martínez-Cayuela, M.; Pérez, J.; Søgaard-Andersen, L.; Muñoz-Dorado, J. Comprehensive Set of Integrative Plasmid Vectors for Copper-Inducible Gene Expression in *Myxococcus xanthus*. *Appl. Environ. Microbiol.* **2012**, *78*, 2515–2521. [CrossRef]
105. Yang, Y.-J.; Singh, R.P.; Lan, X.; Zhang, C.-S.; Li, Y.-Z.; Li, Y.-Q.; Sheng, D.-H. Genome Editing in Model Strain *Myxococcus xanthus* DK1622 by a Site-Specific Cre/loxP Recombination System. *Biomolecules* **2018**, *8*, 137. [CrossRef]
106. Yang, Y.-J.; Wang, Y.; Li, Z.-F.; Gong, Y.; Zhang, P.; Hu, W.-C.; Sheng, D.-H.; Li, Y.-Z. Increasing on-target cleavage efficiency for CRISPR/Cas9-induced large fragment deletion in *Myxococcus xanthus*. *Microb. Cell Factories* **2017**, *16*, 1–15. [CrossRef]
107. Peng, R.; Wang, Y.; Feng, W.-W.; Yue, X.-J.; Chen, J.-H.; Hu, X.-Z.; Li, Z.-F.; Sheng, D.-H.; Zhang, Y.-M.; Li, Y.-Z. CRISPR/dCas9-mediated transcriptional improvement of the biosynthetic gene cluster for the epothilone production in *Myxococcus xanthus*. *Microb. Cell Factories* **2018**, *17*, 1–12. [CrossRef]
108. Whitworth, D.E.; Holmes, A.B.; Irvine, A.G.; Hodgson, D.A.; Scanlan, D.J. Phosphate Acquisition Components of the *Myxococcus xanthus* Pho Regulon Are Regulated by both Phosphate Availability and Development. *J. Bacteriol.* **2008**, *190*, 1997–2003. [CrossRef]
109. Kroos, L.; Kaiser, D. Construction of Tn5 lac, a transposon that fuses lacZ expression to exogenous promoters, and its introduction into *Myxococcus xanthus*. *Proc. Natl. Acad. Sci. USA* **1984**, *81*, 5816–5820. [CrossRef] [PubMed]
110. Kimura, Y.; Mishima, Y.; Nakano, H.; Takegawa, K. An Adenylyl Cyclase, CyaA, of *Myxococcus xanthus* Functions in Signal Transduction during Osmotic Stress. *J. Bacteriol.* **2002**, *184*, 3578–3585. [CrossRef] [PubMed]
111. Diodati, M.E.; Ossa, F.; Caberoy, N.B.; Jose, I.R.; Hiraiwa, W.; Igo, M.M.; Singer, M.; Garza, A.G. Nla18, a Key Regulatory Protein Required for Normal Growth and Development of *Myxococcus xanthus*. *J. Bacteriol.* **2006**, *188*, 1733–1743. [CrossRef] [PubMed]
112. Overgaard, M.; Wegener-Feldbrügge, S.; Søgaard-Andersen, L. The Orphan Response Regulator DigR Is Required for Synthesis of Extracellular Matrix Fibrils in *Myxococcus xanthus*. *J. Bacteriol.* **2006**, *188*, 4384–4394. [CrossRef]
113. Pham, V.D.; Shebelut, C.W.; Jose, I.R.; Hodgson, D.A.; Whitworth, D.E.; Singer, M. The response regulator PhoP4 is required for late developmental events in *Myxococcus xanthus*. *Microbiol.* **2006**, *152*, 1609–1620. [CrossRef] [PubMed]
114. Kadam, S.V.; Wegener-Feldbrügge, S.; Søgaard-Andersen, L.; Velicer, G.J. Novel Transcriptome Patterns Accompany Evolutionary Restoration of Defective Social Development in the Bacterium *Myxococcus xanthus*. *Mol. Biol. Evol.* **2008**, *25*, 1274–1281. [CrossRef]
115. Bode, H.B.; Ring, M.W.; Schwär, G.; Altmeyer, M.O.; Kegler, C.; Jose, I.R.; Singer, M.; Müller, R. Identification of Additional Players in the Alternative Biosynthesis Pathway to Isovaleryl-CoA in the Myxobacterium *Myxococcus xanthus*. *ChemBioChem* **2009**, *10*, 128–140. [CrossRef] [PubMed]
116. Shi, X.; Wegener-Feldbrügge, S.; Huntley, S.; Hamann, N.; Hedderich, R.; Søgaard-Andersen, L. Bioinformatics and Experimental Analysis of Proteins of Two-Component Systems in *Myxococcus xanthus*. *J. Bacteriol.* **2008**, *190*, 613–624. [CrossRef]
117. Müller, F.-D.; Treuner-Lange, A.; Heider, J.; Huntley, S.M.; I Higgs, P. Global transcriptome analysis of spore formation in *Myxococcus xanthus* reveals a locus necessary for cell differentiation. *BMC Genom.* **2010**, *11*, 264. [CrossRef]
118. Furusawa, G.; Dziewanowska, K.; Stone, H.; Settles, M.; Hartzell, P. Global analysis of phase variation in *Myxococcus xanthus*. *Mol. Microbiol.* **2011**, *81*, 784–804. [CrossRef]
119. Croucher, N.; Thomson, N.R. Studying bacterial transcriptomes using RNA-seq. *Curr. Opin. Microbiol.* **2010**, *13*, 619–624. [CrossRef]
120. Zhu, L.-P.; Yue, X.-J.; Han, K.; Li, Z.-F.; Zheng, L.-S.; Yi, X.-N.; Wang, H.-L.; Zhang, Y.-M.; Li, Y.-Z. Allopatric integrations selectively change host transcriptomes, leading to varied expression efficiencies of exotic genes in *Myxococcus xanthus*. *Microb. Cell Factories* **2015**, *14*, 105. [CrossRef]

121. Livingstone, P.; Millard, A.D.; Swain, M.T.; Whitworth, D.E. Transcriptional changes when *Myxococcus xanthus* preys on Escherichia coli suggest myxobacterial predators are constitutively toxic but regulate their feeding. *Microb. Genom.* **2018**, *4*, e000152. [CrossRef]
122. Muñoz-Dorado, J.; Moraleda-Munoz, A.; Torres, F.J.M.; Contreras-Moreno, F.J.; Martin-Cuadrado, A.B.; Schrader, J.M.; I Higgs, P.; Pérez, J. Transcriptome dynamics of the *Myxococcus xanthus* multicellular developmental program. *eLife* **2019**, *8*, 50374. [CrossRef] [PubMed]
123. Sharma, G.; Yao, A.I.; Smaldone, G.T.; Liang, J.; Long, M.; Facciotti, M.T.; Singer, M. Global gene expression analysis of the *Myxococcus xanthus* developmental time course. *Genomics* **2021**, *113*, 120–134. [CrossRef]
124. Whitfield, D.L.; Sharma, G.; Smaldone, G.T.; Singer, M. Peripheral rods: A specialized developmental cell type in *Myxococcus xanthus*. *Genomics* **2020**, *112*, 1588–1597. [CrossRef] [PubMed]
125. Sheng, D.-H.; Wang, Y.; Wu, S.-G.; Duan, R.-Q.; Li, Y.-Z. The Regulation of LexA on UV-Induced SOS Response in *Myxococcus xanthus* Based on Transcriptome Analysis. *J. Microbiol. Biotechnol.* **2021**, *31*, 912–920. [CrossRef]
126. Curtis, P.D.; Atwood, J.; Orlando, R.; Shimkets, L.J. Proteins Associated with the *Myxococcus xanthus* Extracellular Matrix. *J. Bacteriol.* **2007**, *189*, 7634–7642. [CrossRef] [PubMed]
127. Kahnt, J.; Aguiluz, K.; Koch, J.; Treuner-Lange, A.; Konovalova, A.; Huntley, S.; Hoppert, M.; Søgaard-Andersen, L.; Hedderich, R. Profiling the Outer Membrane Proteome during Growth and Development of the Social Bacterium *Myxococcus xanthus* by Selective Biotinylation and Analyses of Outer Membrane Vesicles. *J. Proteome Res.* **2010**, *9*, 5197–5208. [CrossRef]
128. Bhat, S.; Zhu, X.; Patel, R.P.; Orlando, R.; Shimkets, L.J. Identification and Localization of *Myxococcus xanthus* Porins and Lipoproteins. *PLoS ONE* **2011**, *6*, e27475. [CrossRef]
129. Evans, A.G.L.; Davey, H.; Cookson, A.; Currinn, H.; Cooke-Fox, G.; Stanczyk, P.J.; Whitworth, D.E. Predatory activity of *Myxococcus xanthus* outer-membrane vesicles and properties of their hydrolase cargo. *Microbiology* **2012**, *158*, 2742–2752. [CrossRef]
130. Dahl, J.L.; Tengra, F.K.; Dutton, D.; Yan, J.; Andacht, T.M.; Coyne, L.; Windell, V.; Garza, A.G. Identification of Major Sporulation Proteins of *Myxococcus xanthus* Using a Proteomic Approach. *J. Bacteriol.* **2007**, *189*, 3187–3197. [CrossRef]
131. Chao, T.; Kalinowski, J.; Nyalwidhe, J.; Hansmeier, N. Comprehensive proteome profiling of the Fe(III)-reducing myxobacterium *Anaeromyxobacter dehalogenans* 2CP-C during growth with fumarate and ferric citrate. *Proteomics* **2010**, *10*, 1673–1684. [CrossRef]
132. Izzat, S.; Rachid, S.; Ajdidi, A.; El-Nakady, Y.A.; Liu, X.-X.; Ye, B.-C.; Müller, R. The ROK like protein of *Myxococcus xanthus* DK1622 acts as a pleiotropic transcriptional regulator for secondary metabolism. *J. Biotechnol.* **2020**, *311*, 25–34. [CrossRef] [PubMed]
133. Schley, C.; Altmeyer, M.O.; Swart, R.; Müller, R.; Huber, C.G. Proteome Analysis of *Myxococcus xanthus* by Off-Line Two-Dimensional Chromatographic Separation Using Monolithic Poly-(styrene-divinylbenzene) Columns Combined with Ion-Trap Tandem Mass Spectrometry. *J. Proteome Res.* **2006**, *5*, 2760–2768. [CrossRef]
134. Leinenbach, A.; Hartmer, R.; Lubeck, M.; Kneissl, B.; Elnakady, Y.A.; Baessmann, C.; Müller, R.; Huber, C.G. Proteome Analysis of *Sorangium cellulosum* Employing 2D-HPLC-MS/MS and Improved Database Searching Strategies for CID and ETD Fragment Spectra. *J. Proteome Res.* **2009**, *8*, 4350–4361. [CrossRef] [PubMed]
135. Berleman, J.E.; Eallen, S.; Danielewicz, M.A.; Remis, J.P.; Egorur, A.; Ecunha, J.; Hadi, M.Z.; Zusman, D.R.; Northen, T.; Witkowska, H.E.; et al. The lethal cargo of *Myxococcus xanthus* outer membrane vesicles. *Front. Microbiol.* **2014**, *5*, 474. [CrossRef] [PubMed]
136. Whitworth, D.E.; Slade, S.E.; Mironas, A. Composition of distinct sub-proteomes in *Myxococcus xanthus*: Metabolic cost and amino acid availability. *Amino Acids* **2015**, *47*, 2521–2531. [CrossRef] [PubMed]
137. Bolten, C.J.; Heinzle, E.; Müller, R.; Wittmann, C. Investigation of the central carbon metabolism of *Sorangium cellulosum*: Metabolic network reconstruction and quantification of pathway fluxes. *J. Microbiol. Biotechnol.* **2009**, *19*, 23–36. [PubMed]
138. Robinson, M.; Son, B.; Kroos, D.; Kroos, L. Transcription factor MrpC binds to promoter regions of hundreds of developmentally-regulated genes in *Myxococcus xanthus*. *BMC Genom.* **2014**, *15*, 1123. [CrossRef] [PubMed]
139. Dereeper, A.; Guignon, V.; Blanc, G.; Audic, S.; Buffet, S.; Chevenet, F.; Dufayard, J.-F.; Guindon, S.; Lefort, V.; Lescot, M.; et al. Phylogeny.fr: Robust phylogenetic analysis for the non-specialist. *Nucleic Acids Res.* **2008**, *36*, W465–W469. [CrossRef] [PubMed]

MDPI
St. Alban-Anlage 66
4052 Basel
Switzerland
Tel. +41 61 683 77 34
Fax +41 61 302 89 18
www.mdpi.com

Microorganisms Editorial Office
E-mail: microorganisms@mdpi.com
www.mdpi.com/journal/microorganisms

www.ingramcontent.com/pod-product-compliance
Lightning Source LLC
Chambersburg PA
CBHW062036220926
51850CB00010B/2568
9783036539249